Ben Stacy Jerrik (Ed.)

Strategic Automated Command and Control System

Ben Stacy Jerrik (Ed.)

Strategic Automated Command and Control System

Intercontinental ballistic missile, Strategic Air Command, United States Strategic Command

Part Press

Publisher:
Part Press is a trademark of
International Book Market Service Ltd., 17 Rue Meldrum, Beau Bassin, 1713-01 Mauritius
Email: info@bookmarketservice.com
Website: www.bookmarketservice.com

Published in 2012

Printed in: U.S.A., U.K., Germany. This book was not produced in Mauritius.

ISBN: 978-613-6-32921-5

Contents

Articles

References

Strategic_Automated_Command_and_Control_System

The **Strategic Automated Command and Control System** (SACCS) is one of myriad command and control systems used to coordinate the operational functions of United States nuclear forces, specifically intercontinental ballistic missiles and nuclear bombers.

Nomenclature

When originally developed, SACCS stood for "SAC Automated Command and Control System". The end of the Cold War hastened the demise of the Strategic Air Command (SAC) and saw the creation of United States Strategic Command (USSTRATCOM). The "S" in the SACCS acronym was retconned to stand for "Strategic".

Layout

SACCS consisted of three main parts:

1. The Data Display System, which consisted of
   1. The Data Display Generators (film printers, character generators, etc)
   2. The Display Subsystem (projectors, control consoles)
2. The Data Processing Central (or System), which consisted of redundant AN/FSQ-31V computers
3. The Data Communication System, which consisted of
   1. The Electronic Data Transmission Control Center (EDTCC or just TCC)
   2. Encryption/Decryption Subsystem (Crypto)
   3. High Speed Data Transmission Equipment (Modems)

Architecture

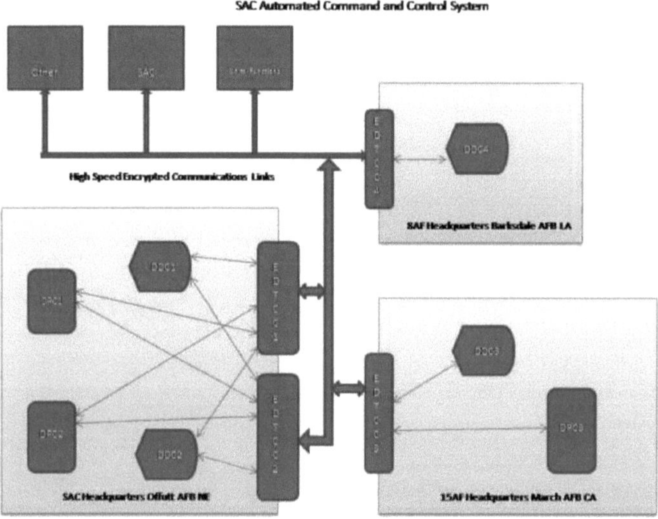

History

The **SACCS** was conceived as an enhancement to the existing manual command, control, and communications and was also an offshoot of the experiences gained by the Air Force and IBM from the development and deployment of the SAGE system. The Data Processing Central element hardware was originally going to be used as the SAGE II computing element, and is architecturally very similar to the SAGE system computer. In 1956 the Commander In Chief of SAC (CINCSAC), who was General Curtis LeMay at the time, determined that SAC's command and control system needed improving, as the current system of leased teleprinter circuits and radio links was too slow to provide realtime communication, which was a necessity during the Cold War. A government program, eventually designated 465L, coordinated military and industry to provide this system. The 465L program was the predecessor to the current Strategic Automated Command Control System. In 1965, SAC procured the 465L system hardware, which was designed to survive nuclear attack and to provide rapid transmission, processing, and display of information to support command and control of SAC's geographically separated forces. The development of the systems software took somewhat longer, as a system of 465L's scope had never been attempted before. By 1970 the system had been for the most part made operational, and was to remain a stable part of SAC C^3 infrastructure until it was replaced by the WWMCCS.[1] [2] [3]

Chronology

- 1958
  - 11 February - Headquarters United States Air Force publishes General Operational Requirement (GOR) #168, for a SAC Automated Command and Control System (SACCS), as the first major documentation of the system[4]
  - 1 April - Headquarters USAF changes the numerical designation of SACCS from Program 133L to 465L.[5]
- 1968
  - 1 January - SACCS attains operational capability[6]
- 1975
  - 6 October - With the replacement of the AN/FSQ-31V computers with Honeywell 6080s, SACCS is officially integrated with Worldwide Military Command and Control System. The AN/FSQ-31V machines were decommissioned in November.

Elements

Maintenance on the SAC HQ SACCS system is performed by the 55th Strategic Communications Squadron, Offutt AFB, Nebraska. Maintenance on the 15th Air Force SACCS system is performed by the 33rd Communications Squadron at March AFB, CA.

Mobile SACCS

There briefly existed a mobile element to SACCS, in the form of a remote communications van. The van was completed on 12 July 1968, and shipped to Andersen AFB, Guam for support to SAC forces. Its current disposition is unknown.[7]

Film

A brief glimse of a **SACCS** Local Communication Center (LCC) mainframe can be seen during the ICBM launch control center scenes in the film "WarGames".

Photos

Part of the SACCS Replacement Keyboard (SRK), Line Printer Unit (LPU) and associated equipment rack is shown at the right edge of this photo depicting an underground missile launch facility

Command Data Buffer configuration

See also

- Post Attack Command and Control System (PACCS)
- Airborne Launch Control System (ALCS)
- Ground Wave Emergency Network (GWEN)
- Minimum Essential Emergency Communications Network (MEECN)
- Survivable Low Frequency Communications System (SLFCS)
- Primary Alerting System (PAS)

References

[1] "Strategic Automated Command Control System" (http://www.globalsecurity.org/wmd/systems/saccs.htm). Global Security.org. . Retrieved December 10, 2006.

[2] "Strategic Automated Command Control System" (http://www.fas.org/nuke/guide/usa/c3i/saccs.htm). Federation of American Scientists. 1999. . Retrieved June 20, 2006.

[3] Wohlman, John (1968). "Computer-Generated Map Data" (http://www.airpower.maxwell.af.mil/airchronicles/aureview/1968/jan-feb/wohlman.html). Air University Review. . Retrieved June 20, 2006.

[4] Strategic Air Command: "Study of SAC Communications System", 6 February 1958

[5] Ibid

[6] Air Force Historical Research Agency: "History of Strategic Air Command: January-June 1968"

[7] Air Force Historical Research Agency: "History of the 3902d Air Base Wing, July - September 1968", pg 40

Intercontinental_ballistic_missile

Test launch of a LGM-25C Titan II ICBM from an underground silo at Vandenberg AFB, during the mid 1970s

An **intercontinental ballistic missile (ICBM)** is a ballistic missile with a long range (greater than 5,500 km or 3,500 miles) typically designed for nuclear weapons delivery (delivering one or more nuclear warheads). Most modern designs support multiple independently targetable reentry vehicles (MIRVs), allowing a single missile to carry several warheads, each of which can strike a different target.

Early ICBMs had limited accuracy that allowed them to be used only against the largest targets such as cities. They were seen as a "safe" basing option, one that would keep the deterrent force close to home where it would be difficult to attack. Attacks against military targets, if desired, still demanded the use of a manned bomber. Second and third generation designs dramatically improved accuracy to the point where even the smallest point targets can be successfully attacked. Similar evolution in size has allowed similar missiles to be placed on submarines, where they are known as submarine launched ballistic missiles, or SLBMs. Submarines are an even safer basing option than land-based missiles, able to move about the ocean at will. This evolution in capability has pushed the manned bomber from the front-line deterrent force in all forces but the United States and Russia, and land-based ICBMs have similarly given way largely to SLBMs.

A Minuteman III ICBM test launch from Vandenberg Air Force Base, California, United States

ICBMs are differentiated by having greater range and speed than other ballistic missiles: intermediate-range ballistic missiles (IRBMs), medium-range ballistic missiles (MRBMs), short-range ballistic missiles (SRBMs)—these shorter range ballistic missiles are known collectively as theatre ballistic missiles. There is no single, standardized definition of what ranges would be categorized as intercontinental, intermediate, medium, or short. Additionally, ICBMs are generally considered to be nuclear only; although several conceptual designs of conventionally armed missiles have been considered, the launch of such a weapon would be such a threat that it would demand a nuclear response, eliminating any military value of such a weapon.

History

World War II

The development of the world's first practical design for an ICBM, A9/10, intended for use in bombing New York and other American cities, was undertaken in Nazi Germany by the team of Wernher von Braun under *Projekt Amerika*. The ICBM A9/A10 rocket initially was intended to be guided by radio, but was changed to be a piloted craft after the failure of Operation Elster. The second stage of the A9/A10 rocket was tested a few times in January and February 1945. The progenitor of the A9/A10 was the German V-2 rocket, also designed by von Braun and widely used at the end of World War II to bomb British and Belgian cities. All of these rockets used liquid propellants. Following the war, von Braun and other leading German scientists were relocated to the United States to work directly for the U.S. Army through Operation Paperclip, developing the IRBMs, ICBMs, and launchers.

This technology was also predicted by US Army General Hap Arnold who wrote in 1943:

> Someday, not too distant, there can come streaking out of somewhere – we won't be able to hear it, it will come so fast – some kind of gadget with an explosive so powerful that one projectile will be able to wipe out completely this city of Washington.[1] [2]

Cold War

The Soviet R-36 (SS-18 Satan) is the largest ICBM in history, with a Throw weight of 8,800 kg, twice that of Peacekeeper.

In the immediate post-war era, the US and USSR both started rocket research programs based on the German wartime designs, especially the V-2. In the US, each branch of the military started its own programs, leading to considerable duplication of effort. In the USSR, rocket research was centrally organized, although several teams worked on different designs. Early designs from both countries were short-range missiles, like the V-2, but improvements quickly followed. China deployed a very small ICBM force of DF-5 missiles beginning in 1981.[3]

The U.S. initiated ICBM research in 1946 with the MX-774 project. This was a three-stage effort with the ICBM development not starting until the third stage. However, funding was cut after only three partially successful launches in 1948 of the second stage design, used to test variations on the V-2 design. With overwhelming air superiority and truly intercontinental bombers, the newly forming US Air Force did not take the problem of ICBM development seriously. Things changed in 1953 with the Soviet testing of their first hydrogen bomb, but it was not until 1954 that the Atlas missile program was given the highest national priority. The Atlas A first flew on 11 June 1957.[4]

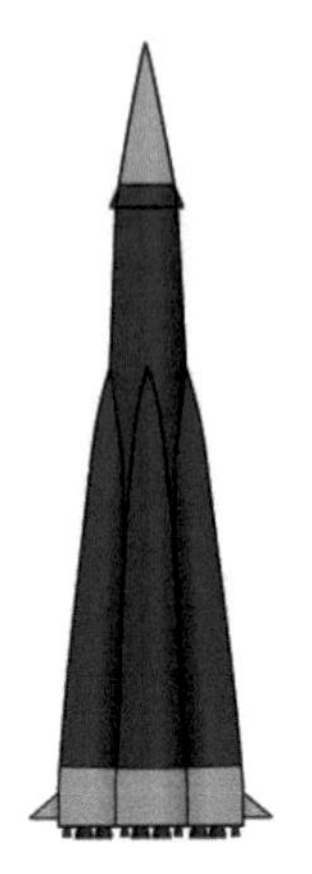

Korolyov's R7 Semyorka Launch configuration.

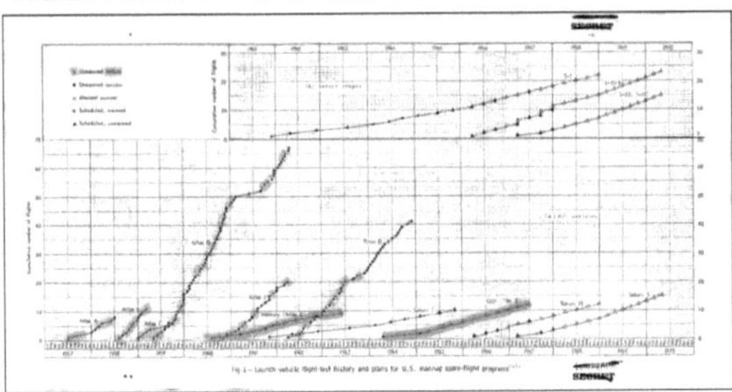

1965 graph of USAF Atlas and Titan ICBM launches, cumulative by month with failures highlighted (pink). This clearly shows how NASA use of ICBM boosters for Projects Mercury and Gemini (blue) served as a highly visible demonstration of confidence in reliability at a time when failure rates had been substantial. (Apollo-Saturn history and projections shown as well.)

The USSR faced different strategic concerns, and early development was focused on missiles able to attack European targets. This changed in 1953 when Sergey Korolyov was directed to start development of a true ICBM able to deliver newly developed hydrogen bombs. Given steady funding throughout, the R-7 developed with some speed, and was successfully tested in August 1957[5] and, on 4 October 1957, placed the first artificial satellite in space, Sputnik. Testing of the R-7 ended in January 1958, but the missile was not considered ready for military service.

The first armed version of the Atlas, the Atlas D, was declared operational in January 1959 at Vandenberg, although it had not yet flown. The first test flight was carried out on 9 July 1959,[6] [7] and the missile was accepted for service on 1 September. Soviet developments quickly followed; the improved R-7A was first flown in December 1959, and declared fully operational in September 1960. The R-7 and Atlas each required a large launch facility, making them vulnerable to attack, and could not be kept in a ready state. Failure rates were very high throughout the early years of ICBM technology. Human spaceflight programs (Vostok, Mercury, Voskhood, Gemini, etc) served as a highly visible means of demonstrating confidence in reliability, with successes translating directly to national defense implications. The US was well behind the Soviet Union in the Space Race, so President Kennedy increased the stakes with the Apollo Program, which used Saturn rocket technology that had been funded by Eisenhower.

U.S. Peacekeeper missile after silo launch.

These early ICBMs also formed the basis of many space launch systems. Examples include Atlas, Redstone, Titan, R-7, and Proton, which was derived from the earlier ICBMs but never deployed as an ICBM. The Eisenhower administration supported the development of solid-fueled missiles such as the LGM-30 Minuteman, Polaris and Skybolt. Modern ICBMs tend to be smaller than their ancestors, due to increased accuracy and smaller and lighter warheads, and use solid fuels, making them less useful as orbital launch vehicles.

The Western view of the deployment of these systems was governed by the strategic theory of Mutual Assured Destruction. In the 1950s and 1960s, development began on Anti-Ballistic Missile systems by both the U.S. and USSR; these systems were restricted by the 1972 ABM treaty. The first successful ABM test were conducted by the USSR in 1961, that later deployed a fully operating system defending Moscow in the 1970s (see Moscow ABM system).

The 1972 SALT treaty froze the number of ICBM launchers of both the USA and the USSR at existing levels, and allowed new submarine-based SLBM launchers only if an equal number of land-based ICBM launchers were dismantled. Subsequent talks, called SALT II, were held from 1972 to 1979 and actually reduced the number of nuclear warheads held by the USA and USSR. SALT II was never ratified by the United States Senate, but its terms were nevertheless honored by both sides until 1986, when the Reagan administration "withdrew" after accusing the USSR of violating the pact.

In the 1980s, President Ronald Reagan launched the Strategic Defense Initiative as well as the MX and Midgetman ICBM programs.

China developed a minimal independent nuclear deterrent entering its own cold war after an idealogical split with the Soviet Union beginning in the early 1960s. After first testing a domestic built nuclear weapon in 1964, it went on to develop various warheads and missiles. Beginning in the early 1970s, the liquid fuelled DF-5 ICBM was developed and used as a satellite launch vehicle in 1975. The DF-5, with range of 10,000 to 12,000 km long enough to strike the western US and the USSR, was silo deployed with the first pair in service by 1981 with possibly twenty missiles in service by the late 1990s.[8] China also deployed the JL-1 Medium-range ballistic missile with a reach of 1700 km aboard the ultimately unsuccessful type 92 submarine.[9]

Post–Cold War

In 1991, the United States and the Soviet Union agreed in the START I treaty to reduce their deployed ICBMs and attributed warheads.

As of 2009, all five of the nations with permanent seats on the United Nations Security Council have operational ICBM systems: all have submarine-launched missiles, and Russia, the United States and China also have land-based missiles. In addition, Russia and China have mobile land-based missiles.

Israel is believed to have deployed a road mobile nuclear ICBM, the Jericho III, which entered service in 2008, an upgraded version is in development.[10] [11]

India successfully test fired Agni V, with a strike range of more than 5,000 km on 19th April 2012, claiming entry into the ICBM club.[12] .

It is speculated by some intelligence agencies that North Korea is developing an ICBM;[13] two tests of somewhat different developmental missiles in 1998 and 2006 were not fully successful.[14] [15] On 5 April 2009, North Korea launched a missile. They claimed that it was to launch a satellite, but there is no proof to back up that claim.[16] Likewise, North Korea attempted another test fire in April 2012, claimed also as a satellite launch, but it broke up in flight after 90 seconds.

Most countries in the early stages of developing ICBMs have used liquid propellants, with the known exceptions being the Indian Agni-V, the planned South African RSA-4 ICBM and the now in service Israeli Jericho 3.[17]

Flight phases

The following flight phases can be distinguished:

- boost phase: 3 to 5 minutes (shorter for a solid rocket than for a liquid-propellant rocket); altitude at the end of this phase is typically 150 to 400 km depending on the trajectory chosen, typical burnout speed is 7 km/s, up to the speed of Low Earth Orbit.
- midcourse phase: approx. 25 minutes—sub-orbital spaceflight in an elliptic flightpath; the flightpath is part of an ellipse with a vertical major axis; the apogee (halfway through the midcourse phase) is at an altitude of approximately 1,200 km; the semi-major axis is between 3,186 km and 6,372 km; the projection of the flightpath on the Earth's surface is close to a great circle, slightly displaced due to earth rotation during the time of flight; the missile may release several independent warheads, and penetration aids such as metallic-coated balloons, aluminum chaff, and full-scale warhead decoys.
- reentry phase (starting at an altitude of 100 km): 2 minutes – impact is at a speed of up to 4 km/s (for early ICBMs less than 1 km/s); see also maneuverable reentry vehicle.

Modern ICBMs

Modern ICBMs typically carry multiple independently targetable reentry vehicles (*MIRVs*), each of which carries a separate nuclear warhead, allowing a single missile to hit multiple targets. MIRV was an outgrowth of the rapidly shrinking size and weight of modern warheads and the Strategic Arms Limitation Treaties which imposed limitations on the number of launch vehicles (SALT I and SALT II). It has also proved to be an "easy answer" to proposed deployments of ABM systems—it is far less expensive to add more warheads to an existing missile system than to build an ABM system capable of shooting down the additional warheads; hence, most ABM system proposals have been judged to be impractical. The first operational ABM systems were deployed in the U.S. during 1970s. Safeguard ABM facility was located in North Dakota and was operational from 1975–1976. The USSR deployed its Galosh ABM system around Moscow in the 1970s, which remains in service. Israel deployed a national ABM system based on the Arrow missile in 1998,[18] but it is mainly designed to intercept shorter-ranged theater ballistic missiles, not ICBMs. The U.S. Alaska-based National missile defense system attained initial operational capability in 2004.[19]

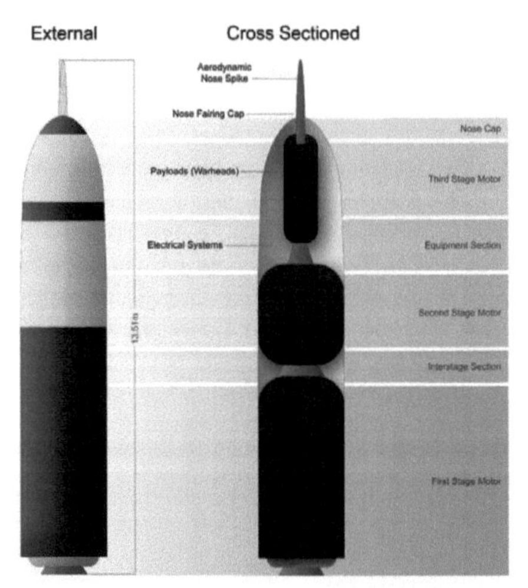

External and cross sectional views of a Trident II D5 nuclear missile system. It is a submarine launched missile capable of carrying multiple nuclear warheads up to 8,000 km. Trident missiles are carried by fourteen active US Navy Ohio class submarines and four Royal Navy Vanguard class submarines.

ICBMs can be deployed from multiple platforms:

- in missile silos, which offer some protection from military attack (including, the designers hope, some protection from a nuclear first strike)
- on submarines: submarine-launched ballistic missiles (SLBMs); most or all SLBMs have the long range of ICBMs (as opposed to IRBMs)
- on heavy trucks; this applies to one version of the RT-2UTTH Topol M which may be deployed from a self-propelled mobile launcher, capable of moving through roadless terrain, and launching a missile from any point along its route
- mobile launchers on rails; this applies, for example, to РТ-23УТТХ "Молодец" (RT-23UTTH "Molodets"—SS-24 "Scalpel")

The last three kinds are mobile and therefore hard to find.

During storage, one of the most important features of the missile is its serviceability. One of the key features of the first computer-controlled ICBM, the Minuteman missile, was that it could quickly and easily use its computer to test itself.

In flight, a booster pushes the warhead and then falls away. Most modern boosters are solid-fueled rocket motors, which can be stored easily for long periods of time. Early missiles used liquid-fueled rocket motors. Many liquid-fueled ICBMs could not be kept fuelled all the time as the cryogenic liquid oxygen boiled off and caused ice formation, and therefore fueling the rocket was necessary before launch. This procedure was a source of significant operational delay, and might allow the missiles to be destroyed by enemy counterparts before they could be used. To resolve this problem the British invented the missile silo that protected the missile from a first strike and also hid fuelling operations underground.

Once the booster falls away, the warhead continues on an unpowered ballistic trajectory, much like an artillery shell or cannon ball. The warhead is encased in a cone-shaped reentry vehicle and is difficult to detect in this phase of flight as there is no rocket exhaust or other emissions to mark its position to defenders. The high speeds of the warheads make them difficult to intercept and allow for little warning striking targets many thousands of kilometers away from the launch site (and due to the possible locations of the submarines: anywhere in the world) within approximately 30 minutes.

Many authorities say that missiles also release aluminized balloons, electronic noisemakers, and other items intended to confuse interception devices and radars (see penetration aid).

As the nuclear warhead reenters the Earth's atmosphere its high speed causes friction with the air, leading to a dramatic rise in temperature which would destroy it if it were not shielded in some way. As a result, warhead components are contained within an aluminium honeycomb substructure, sheathed in pyrolytic graphite-epoxy resin composite, with a heat-shield layer on top which is constructed out of 3-Dimensional Quartz Phenolic.

Accuracy is crucial, because doubling the accuracy decreases the needed warhead energy by a factor of four. Accuracy is limited by the accuracy of the navigation system and the available geophysical information.

Strategic missile systems are thought to use custom integrated circuits designed to calculate navigational differential equations thousands to millions of times per second in order to reduce navigational errors caused by calculation alone. These circuits are usually a network of binary addition circuits that continually recalculate the missile's position. The inputs to the navigation circuit are set by a general purpose computer according to a navigational input schedule loaded into the missile before launch.

One particular weapon developed by the Soviet Union (FOBS) had a partial orbital trajectory, and unlike most ICBMs its target could not be deduced from its orbital flight path. It was decommissioned in compliance with arms control agreements, which address the maximum range of ICBMs and prohibit orbital or fractional-orbital weapons.

Low-flying guided cruise missiles are an alternative to ballistic missiles.

Specific missiles

Land-based ICBMs

Russia, the United States and China are the only countries currently known to possess land-based ICBMs.[20]

Testing of the Peacekeeper re-entry vehicles at the Kwajalein Atoll. All eight fired from only one missile. Each line, if its warhead were live, represents the potential explosive power of about 375 kilotons of TNT, about twelve times larger than the detonation of atomic bomb in Hiroshima.

The United States currently operates 450 ICBMs in three USAF bases. The only model deployed is LGM-30G Minuteman-III.

All previous USAF Minuteman II missiles have been destroyed in accordance with START, and their launch silos have been sealed or sold to the public. To comply with the START II most U.S. multiple independently targetable reentry vehicles, or MIRVs, have been eliminated and replaced with single warhead missiles. The powerful MIRV-capable Peacekeeper missiles were phased out in 2005.[21] However, since the abandonment of the START II treaty, the U.S. is said to be considering retaining 800 warheads on an existing 450 missiles.[22]

The Russian Strategic Rocket Forces have 369 ICBMs able to deliver 1,247 nuclear warheads, 58 silo-based R-36M2 (SS-18), 70 silo-based UR-100N (SS-19), 171 mobile RT-2PM "Topol" (SS-25), 52 silo-based RT-2UTTH "Topol M" (SS-27), 18 mobile RT-2UTTH "Topol M" (SS-27), 6 (15 in December 2011[23]) mobile RS-24 "Yars" (SS-29) *(Future replacement for R-36 & UR-100N missiles)*

China has developed several long range ICBMs, like the DF-31. The Dongfeng 5 or DF-5 is a 3 stage liquid fuel ICBM and has an estimated range of 13,000 kilometers. The DF-5 had its first flight in 1971 and was in operational service 10 years later. One of the downsides of the missile was that it took between 30 and 60 minutes to fuel. The Dong Feng 31 (a.k.a. CSS-10) is a medium-range, three stage, solid propellant intercontinental ballistic missile, and is a land-based variant of the submarine launched JL-2. The DF-41 or CSS-X-10 can carry up to 10 nuclear warheads, which are maneuverable reentry vehicles and has a range of approximately 12,000–14,000 km.[24] [25] [26] [27]

Israel is believed to have deployed a road mobile nuclear ICBM, the Jericho III, which entered service in 2008. It is possible for the missile to be equipped with a single 750 kg nuclear warhead or up to three MIRV warheads. It is believed to be based on the Shavit space launch vehicle and is estimated to have a range of 4,800 to 11,500 km[10] (2,982 to 7,180 miles). In November 2011 Israel tested an ICBM believed to be an upgraded version of the Jericho III.[11]

India has a series of ballistic missiles called Agni, of which the latest is Agni-V. On 19 April 2012, India successfully test fired Agni-V, a three stage solid fueled missile, with a strike range of more than 5,000 km.[28]

Submarine-launched

All current designs of submarine launched ballistic missiles have intercontinental range. Current operators of such missiles are the United States, Russia, United Kingdom, and France. The People's Republic of China and India are both working on near term deployable SLBM systems; although from 1986, China had deployed a system from the Type 092 submarine. One of the two submarines was lost at sea, and neither of the ultimately unsuccessful class was ever believed to have deployed away from home waters.[29]

See also

Artist's concept of SS-24 deployed on railway.

- Air Force Space Command
- Anti-ballistic missile
- Anti-Ballistic Missile Treaty
- Atmospheric reentry
- Countermeasure
- Dense Pack
- Fractional Orbital Bombardment System
- France and weapons of mass destruction
- General Bernard Adolph Schriever
- Heavy ICBM
- High-alert nuclear weapon
- People's Republic of China and weapons of mass destruction
- India and weapons of mass destruction
- Israel and weapons of mass destruction

- List of ICBMs
- Missile Defense Agency
- Nuclear disarmament
- Nuclear navy
- Nuclear warfare
- Nuclear weapon
- Russia and weapons of mass destruction
- SLBM
- Strike Force (France)
- Submarine
- Throw-weight
- United Kingdom and weapons of mass destruction
- United States and weapons of mass destruction

References

[1] "Toward a Theory of Space Power" (http://www.ndu.edu/press/space-Ch19.html). Ndu.edu. . Retrieved 2012-04-19.
[2] Correll, John T.. "How the Air Force Got the ICBM" (http://www.airforce-magazine.com/MagazineArchive/Pages/2005/July 2005/0705icbm.aspx). Airforce-magazine.com. . Retrieved 2012-04-19.
[3] https://www.fas.org/nuke/guide/china/icbm/df-5.htm
[4] "Atlas" (http://www.century-of-flight.net/Aviation history/space/Atlas.htm), Century of Flight
[5] Wade, Mark. "R-7" (http://www.astronautix.com/lvs/r7.htm). *Encyclopedia Astronautica*. . Retrieved 4 July 2011.
[6] "Atlas D" (http://www.missilethreat.com/missilesoftheworld/id.15/missile_detail.asp). Missile Threat. . Retrieved 19 April 2012.
[7] "Encyclopedia Astronautica: Atlas" (http://www.astronautix.com/lvs/atlas.htm). Astronautix.com. . Retrieved 19 April 2012.
[8] https://www.fas.org/nuke/guide/china/icbm/df-5.htm
[9] https://www.fas.org/nuke/guide/china/slbm/type_92.htm
[10] Andrew Feickert (5 March 2004). *"Missile Survey: Ballistic and Cruise Missiles of Foreign Countries"* (http://www.au.af.mil/au/awc/awcgate/crs/rl30427.pdf). *Congressional Research Service* ~(The Library of Congress). RL30427. . Retrieved 21 June 2010.
[11] Pfeffer, Anshel (2 November 2011). "IDF test-fires ballistic missile in central Israel" (http://www.haaretz.com/news/diplomacy-defense/idf-test-fires-ballistic-missile-in-central-israel-1.393306). *Haaretz*. . Retrieved 3 November 2011.
[12] "News / National : Agni-V successfully test-fired" (http://www.thehindu.com/news/national/article3330921.ece). *The Hindu*. 19 April 2012. . Retrieved 19 April 2012.
[13] "Taep'o-dong 2 (TD-2) – North Korea" (http://www.fas.org/nuke/guide/dprk/missile/td-2.htm). Fas.org. . Retrieved 19 April 2012.
[14] "CNN.com" (http://edition.cnn.com/2006/WORLD/asiapcf/07/04/korea.missile/). CNN. . Retrieved 19 April 2012.
[15] CNN.com (http://www.cnn.com/2006/WORLD/asiapcf/07/05/korea.missile/index.html)
[16] "BBC.co.uk" (http://news.bbc.co.uk/2/hi/asia-pacific/7982874.stm). BBC News. 5 April 2009. . Retrieved 19 April 2012.
[17] Astronautix.com (http://www.astronautix.com/lvfam/jericho.htm)
[18] "Israeli Arrow ABM System is Operational as War Clouds Darken" (http://www.ishitech.co.il/1102ar1.htm). Ishitech.co.il. . Retrieved 19 April 2012.
[19] "MissileThreat.com" (http://www.missilethreat.com/systems/fort_greely.html). MissileThreat.com. 8 December 1998. . Retrieved 19 April 2012.
[20] "Britannica.com" (http://www.britannica.com/EBchecked/topic/290047/ICBM). Britannica.com. 3 May 1961. . Retrieved 19 April 2012.
[21] This story was written by 2nd Lt. Joshua S. Edwards (20 September 2005). "Peacekeeper missile mission ends during ceremony" (http://www.af.mil/news/story.asp?storyID=123011845). Af.mil. . Retrieved 19 April 2012.
[22] Nuclear Notebook: U.S. and Soviet/Russian intercontinental ballistic missiles, 1959–2008 (http://thebulletin.metapress.com/content/037qk4866n431045/fulltext.pdf) *Bulletin of the Atomic Scientists* (http://thebulletin.org/), January/February 2009
[23] Second RS-24 regiment begins combat duty. "Second RS-24 regiment begins combat duty – Blog – Russian strategic nuclear forces" (http://russianforces.org/blog/2011/12/second_rs-24_regiment_begins_c.shtml). Russianforces.org. . Retrieved 19 April 2012.
[24] "Five types of missiles to debut on National Day_English_Xinhua" (http://news.xinhuanet.com/english/2009-09/02/content_11982723.htm). News.xinhuanet.com. 2 September 2009. . Retrieved 6 April 2010.
[25] John Pike. "DF-41 – China Nuclear Forces" (http://www.globalsecurity.org/wmd/world/china/df-41.htm). Globalsecurity.org. . Retrieved 6 April 2010.
[26] "DF-41 (CSS-X-10) (China) – Jane's Strategic Weapon Systems" (http://www.janes.com/articles/Janes-Strategic-Weapon-Systems/DF-41-CSS-X10-China.html). Janes.com. 2 July 2009. . Retrieved 6 April 2010.
[27] "CSS-X-10 (DF-41)" (http://www.missilethreat.com/missilesoftheworld/id.35/missile_detail.asp). MissileThreat. . Retrieved 6 April 2010.
[28] "News / National : Agni-V successfully test-fired" (http://www.thehindu.com/news/national/article3330921.ece). *The Hindu*. 19 April 2012. . Retrieved 19 April 2012.
[29] John Pike (24 July 2011). "Type 092 Xia Class SSBN" (http://www.globalsecurity.org/wmd/world/china/type_92.htm). Globalsecurity.org. . Retrieved 19 April 2012.

External links

- Ballistic missile characters (http://www.mda.mil/mdalink/bcmt/bm_char_1.htm)
- Estimated Strategic Nuclear Weapons Inventories (September 2004) (http://es.rice.edu/projects/Poli378/Nuclear/f04.stratg_invent.html)
- Intercontinental Ballistic and Cruise Missiles (http://www.fas.org/nuke/guide/usa/icbm/index.html)
- "A Tale of Two Airplanes" (http://www.RC135.com/) by Kingdon R. "King" Hawes, Lt Col, USAF (Ret.)

Strategic_Air_Command

Strategic Air Command	
Strategic Air Command emblem	
Active	1946 - 1992
Country	USA
Branch	US Army Air Forces (1946-1947) US Air Force (1947-1992)
Type	Major Command
Garrison/HQ	Offutt Air Force Base, Nebraska
Motto	"Peace is our Profession"
Commanders	
Notable commanders	Curtis LeMay

The **Strategic Air Command** (**SAC**) was both a Major Command (MAJCOM) of the United States Air Force and a "specified command" of the United States Department of Defense. SAC was the operational establishment in charge of America's land-based strategic bomber aircraft and land-based intercontinental ballistic missile (ICBM) strategic nuclear arsenal from 1946 to 1992. SAC also controlled the infrastructure necessary to support the strategic bomber and ICBM operations, such as aerial refueling tanker aircraft to refuel the bombers in flight, strategic reconnaissance aircraft, command post aircraft, and, until 1957, fighter escorts.

Following the fall of the Soviet Union, the Air Force instituted a comprehensive reorganization of its major commands. As part of this reorganization, SAC was disestablished on 1 June 1992. As part of the reorganization, SAC's bomber aircraft, ICBMs, strategic reconnaissance aircraft, and command post aircraft were merged with USAF fighter and other tactical aircraft assets and reassigned to the newly-established Air Combat Command (ACC). This included B-52 and B-1 bomber aircraft assigned to the Air Force Reserve and Air National Guard, respectively.

At the same time, most of SAC's aerial refueling tanker aircraft, including those in the Air Force Reserve and Air National Guard, were reassigned to the new Air Mobility Command (AMC). Tankers based in Europe were reassigned to United States Air Forces in Europe (USAFE), while regular air force tankers in the Pacific, as well as Alaska Air National Guard tankers, were reassigned to Pacific Air Forces (PACAF).

The ICBM force was later transferred from ACC to the Air Force Space Command (AFSPC) on 1 July 1993. Another change in late 2009 and early 2010 resulted in the transfer of the ICBM force from AFSPC and the B-52 and B-2 strategic bomber force from ACC to the newly-established Air Force Global Strike Command (AFGSC), which is a direct descendant of SAC.[1]

History

Early history 1946–58

Special photo of Air Force bombers from the 1930s through the late 1940s. A Douglas B-18 "Bolo"; a Boeing B-17 "Flying Fortress"; a Boeing "B-29 Superfortress" and the B-36 "Peacemaker" dominating the group photo with a 230 Ft Wingspan. Taken at Carswell AFB, Texas after the receipt of the first B-36 in 1948. Note the SAC 7th Bombardment Wing marking on the B-29.

During the interwar period between World War I and World War II, a group of U.S. Army Air Corps officers colloquially referred to as the Bomber Mafia, convinced of the potential of strategic bombing, paved the way both for the massive strategic air campaigns in Europe and the Pacific in World War II and the later creation of SAC. SAC's United States Army Air Forces predecessor, the Continental Air Forces, was established on 13 December 1944 and activated on 15 December 1944. CAF controlled the numbered air forces within the United States (1st Air Force, 2nd Air Force, 3rd Air Force and 4th Air Force) and their training mission.

On 21 March 1946, CAF was disestablished as part of a major reorganization of the USAAF. Within the United States, the USAAF was divided into three separate commands: Tactical Air Command (TAC), Air Defense Command (ADC), and Strategic Air Command (SAC). Airfields formerly assigned to CAF were reassigned to one of these three major commands.

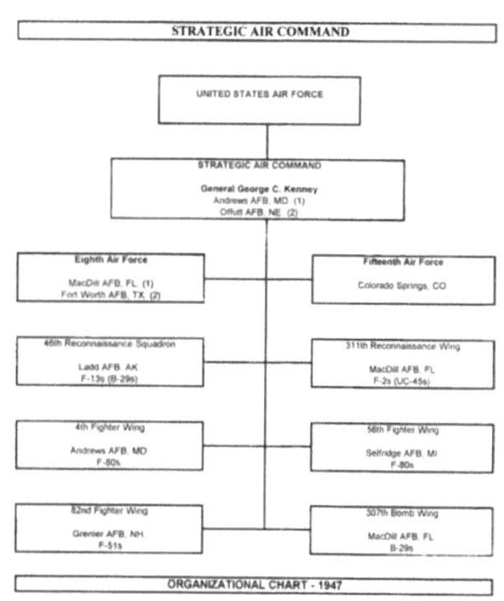

SAC organisation 1947. Source: Ron Mixer, "The Genealogy of the Strategic Air Command", Battermix.

SAC's original headquarters was located at Bolling Field in Washington, DC, the headquarters of the disestablished Continental Air Forces, with the headquarters organization of CAF being redesignated as Strategic Air Command. Its first commander was General George C. Kenney.[2] :29–30 Ten days later, Fifteenth Air Force was assigned to the command as its first Numbered Air Force. There were thirteen bombardment groups assigned to Continental Air Forces just before its redesignation as SAC. These included the 40th (effectively became 43rd), 44th, the 93rd, 444th, 448th (became 92nd), 449th, 467th (effectively became 301st), 485th, and 498th (became 307th). There was also the 58th Bombardment Wing, Very Heavy, which supervised the *Silverplate* atomic-capable 509th Composite Group.[3] Also active was the 73rd Bombardment Wing, Very Heavy, transferred from Third Air Force. However several of these units were quickly disbanded, or renumbered to preserve the heritage of other units. In June 1946 Eighth Air Force was also assigned. SAC HQ then moved to Andrews AFB, MD on 20 October 1946.

Strategic Air Command was created with the stated mission of providing long range bombing capabilities anywhere in the world. But due to the massive post-World War II demobilization of the U.S. armed forces, Kenney's position at the UN Military Staff Committee in New York, and Kenney's unhappiness with being assigned to SAC, for the first two years of its existence, there was some lack of urgency. Kenney's deputy, Maj. Gen. St. Clair Streett, wrote in July 1946: "No major strategic threat or requirement now exists, in the opinion of our country's best strategists nor will such a requirement exist for the next three to five years."[4]

The situation began to change on 19 October 1948, when Lieutenant General Curtis LeMay assumed leadership of the Strategic Air Command, a position he held until June 1957, the longest tenure for any United States armed forces commander since Winfield Scott.[2] :99 Soon after taking command, on 9 November, LeMay relocated SAC to Offutt AFB south of Omaha. It was under the leadership of LeMay that SAC developed the technical capability, strategic planning, and operational readiness to carry out its strategic mission anywhere in the world. Among the technological developments that made this possible were the widescale use of in-flight refueling, jet engines, and intercontinental ballistic missiles.

The development of jet aircraft, specifically the B-47 Stratojet, was a key component in building the Strategic Air Command's bombing capacity. When LeMay assumed command of SAC, his vision was to create a force of nuclear-armed long-range bombers with the capability to devastate the Soviet Union within a few days of the advent of war.[2] :102 But the reality when LeMay assumed command was that SAC had only sixty nuclear capable aircraft, none of which had the long-range capabilities he desired.[5]

Introduced into active service in 1951, the B-47 was the first jet aircraft employed by SAC. Despite having a limited range, by the end of LeMay's command in 1957, the B-47 had become the backbone of SAC, comprising over half of its total aircraft and eighty percent of its bomber capacity.[2] :104 A key factor enabling the B-47 to become the mainstay of SAC (and to fulfill LeMay's desire for a long range bomber) was the development of in-flight refueling. In addition, "Reflex" operations based in forward countries such as Morocco, Spain and Turkey provided infra-structure for temporary duty (TDY) assignment of US-based B-47 bomb wings. Sixteenth Air Force managed SAC operations in Morocco and Spain from 1957 to 1966.

In-flght refueling, long a dream of airmen, became a reality in 1954 with the introduction of the KC-97 Stratotanker into active service. The primary reason it became essential to SAC was the limited range of about 2000 miles of the B-47.[5] :108 In-flight fueling gave the B-47 unlimited range and the ability to fly for extended periods of time. This new ability was openly demonstrated to the USSR with several well publicized non-stop flights around the world. The development meant that SAC was no longer dependent on stationing nuclear capable bombers in foreign countries like Spain and Britain, which proved to be politically sensitive in the late 1940s/early 1950s.[5] :108

Along with in-flight refueling, another important element in the growth of SAC was the development of ballistic missiles. The rapid development of ballistic missiles in the 1950s provided SAC with another means of carrying out its mission of being able to strike anywhere in the world. While the U.S. Air Force had started a missile development program in 1946, it was not seriously pursued until reports surfaced about the progress of Soviet Union rocket technology and the threat it posed to the US.[2] :112–13 The perceived threat motivated the Eisenhower administration to make ballistic missiles a top priority and tasked Air Force Brigadier General Bernard Schriever with leading the development program. By 1958, roughly four years after Schriever had initiated his ballistic missile program, SAC activated the 704th Strategic Missile Wing to operate first the intermediate range Thor missile and then a year later the first true ICBM, the Atlas missile.[2] :117–18 Schriever followed up his quick development of the two missile systems with the development of the Titan II and Minuteman missile systems shortly thereafter.

From 1946/47 to 1957, SAC also incorporated fighter escort wings and later strategic fighter wings. Intended to escort bombers to their targets in a continuation of World War II practice, they were equipped with F-51s and later F-84s. There were a total of ten. Eighth Air Force was assigned the 12th, 27th, and 33rd Wings, and Fifteenth Air Force the 56th, 71st, 82nd, 407th Wings. They were phased out in 1957-58.[6]

During LeMay's command, SAC was able to effect great changes in American nuclear strategy. At the beginning of the Cold War, SAC was effectively powerless in shaping the American nuclear strategy it was tasked with carrying out. The four main issues instrumental in forming the nuclear strategy were technical limitations, nuclear weapon availability, lack of strategic thinking and politics. The first of the two factors of technical limitations and availability went hand in hand, as from 1946–48, the US had only twelve atomic bombs and between five and twenty-seven B-29s capable of delivering the bombs.[7] The lack of strategic thinking was largely a result of the unfamiliarity of the atomic bomb and the high level of secrecy with which it had been developed. That began to change in 1948 when reports of Bikini Atoll tests were circulated among the Air Force, which made information about the bomb more available to planners and helped to convince them of its strategic capabilities.[7] :67 The new strategic thinking found its place in the proposed Joint Emergency War Plan codenamed "Halfmoon", which called for the dropping of fifty atomic bombs on twenty cities in the Soviet Union.[7] :68 At this point, politics entered into the formation of nuclear strategy in the form of President Harry S. Truman. The president initially rejected "Halfmoon" and ordered the development of a non-nuclear alternative plan, only to later change his mind during the Berlin Blockade.[7] :68–9 These four factors combined to create a high level of uncertainty and prevented the development of an effective nuclear strategy.

It was this uncertainty that LeMay entered into upon assuming command of SAC which emboldened him and SAC planners to attempt to unilaterally form American nuclear strategy. LeMay started shortly after his arrival at SAC, by having SAC planners draw up Emergency War Plan 1-49, which involved striking seventy Soviet cities with 133 atomic bombs over a thirty day period in an effort to destroy Soviet industrial capacity.[7] But with the Soviet Union gaining possession of atomic weapons in 1949, SAC was forced to rethink its nuclear strategy. Under orders from the Joint Chiefs of Staff, SAC was told its primary objective was bombing targets in order to damage or destroy Soviet ability to deliver nuclear weapons, its secondary objective was stopping Soviet advances into Western Europe, and its tertiary objective was the same as before, destroying Soviet industrial capacity.[8] The redefinition and expansion of its mission would help SAC to formalize and consolidate its control over nuclear planning and strategy. This was done by LeMay in a 1951 meeting with high level Air Force staff, when he convinced them that unreasonable operational demands were being placed on SAC and, in order to alleviate the issue, SAC should be allowed to approve target selections before they were finalized.[8] :18

SAC's assumption of control over nuclear strategy led to the adoption of a strategy based on the idea of counterforce. SAC planners understood that as the Soviet Union increased their nuclear capacity, destroying or "countering" those forces (bombers, missiles, etc.) became of greater strategic importance than destroying industrial capacity.[5] :100 In 1954, the Eisenhower administration concurred with the new focus, with the President expressing a preference for military over civilian targets.[8] :35 While the Eisenhower administration approved of the strategy in general, LeMay continued to increase SAC's independence by refusing to submit SAC war plans for review, believing that operational plans should be closely guarded, a view the Joint Chiefs of Staff eventually came to accept.[8] :37 By the end of the 1950s, SAC had identified 20,000 potential Soviet target sites and had officially designated 3,560 of those sites as bombing targets, with the significant percentage being counterforce targets of Soviet air defense, airfields and suspected missile sites.[8] :60 LeMay and SAC's continuing efforts to assume greater control over nuclear strategy were vindicated on August 11, 1960, when Eisenhower approved a plan to create the Joint Strategic Target Planning Staff (dominated by SAC) to prepare the National Strategic Target List and the Single Integrated Operation Plan (SIOP) for nuclear war.[8] :62

Besides developing and implementing new technology and strategies, SAC was actively involved in the Korean War. Shortly after beginning of the war in July 1950, SAC dispatched ten nuclear-capable bombers to Andersen AFB on Guam under orders from the Joint Chiefs.[5] :112 But SAC did more than just provide a nuclear option during the Korean War, It also deployed four B-29 bomber wings that were used in tactical operations against enemy forces and logistics[5] :114 All of this led LeMay to express concern that "too many splinters were being whittled off the stick", preventing him from being able to carry out his primary mission of strategic deterrence.[5] :113–4 As a result, LeMay was relieved when the Korean War ended in 1953 and he was able to go back to building SAC's arsenal and gaining control over nuclear strategy.

Height of the Cold War 1959-1990

Original SAC patch

The late 1950s and early 1960s heralded the arrival of two new bomber aircraft, the supersonic B-58 Hustler and B-70 Valkyrie. Both were intended to use a high-speed, high-altitude bombing approach that followed a trend of bombers flying progressively faster and higher since the start of manned bomber use. However, the 1960 shootdown of a CIA U-2 by an SA-2 Guideline surface-to-air missile (SAM) while on a clandestine reconnaissance mission over the Soviet Union suddenly complicated the introduction of these new aircraft. SAC now found itself in an uncomfortable position; new bombers which had been tuned for efficiency at high speeds and altitudes, performance that had been purchased at great cost. The B-70 was intended to eventually replace the B-52 in the long-range role, and while SAC initially thought it could eliminate the medium bomber role of the B-47, it had retrenched from this position and introduced the B-58 to replace the B-47 in the medium range role. However, the B-58 was expensive to develop and purchase, required enormous amounts of fuel and maintenance in comparison to the B-47, and was estimated to cost three times as much to operate than the much larger and longer-ranged B-52. Flying at lower altitudes, where the air density is much higher, drag on the B-58 was significantly higher and limited its range and speed. At low altitudes, the B-58 ended up with performance that was only a small improvement over the B-47 it was meant to replace. The net result was that the B-70 never entered operational service with SAC and the B-58 only served for 10 years until its withdrawal from service in 1970 and replacement by the FB-111A.

Despite SAC's establishment of "hardened" underground command and control facilities at its headquarters at Offutt AFB, LeMay and his planners knew that a direct nuclear strike by Soviet forces employing hydrogen weapons would likely destroy the facility. As a backup to this potentiality, the concept of a SAC airborne command post was developed. As envisioned, the airborne command post would be carried on a long range/long endurance aircraft, manned by a battle staff headed by a SAC general officer of at least brigadier general rank. The aircraft would be equipped with the latest in electronics and communications equipment so that it would be able to assume control of all of SAC's bomber, aerial refueling and reconnaissance aircraft, as well as SAC's land-based intercontiental ballistic missile (ICBM) force and the U.S. Navy's Fleet Ballistic Missile (FBM) submarine force in the event SAC headquarters was destroyed. Like the B-52, the airborne command post would also be hardened against electromagnetic pulse (EMP) radiation, making it capable of operating during a nuclear exchange with the Soviet Union. A fleet of these aircraft would also enable SAC to keep one such aircraft continuously airborne, 24 hours a day every day of the year. The aircraft selected for this duty was a derivative of SAC's KC-135 Stratotanker. Named the EC-135 Looking Glass, it realized the SAC vision of a flying command post. As a result, one of SAC's EC-135 Looking Glass aircraft was constantly airborne from 1961 until the dissoultion of the Soviet Union and the *de facto* end of the Cold War in 1990.[9]

With the end of the war in Vietnam, SAC refocused its efforts back to deterrence of a nuclear attack by the Soviet Union. Concurrently, the costs of the Vietnam War took a heavy stateside toll on SAC as many of its bases were either deactivated, transferred to other Air Force MAJCOMs, or transferred to other U.S. military services as part of cost-cutting moves during or shortly after the end of combat operations in Southeast Asia. Older B-52B, B-52C, B-52E and B-52F aircraft were retired, along with the B-58A, leaving SAC with an offensive force of several hundred B-52D, B-52G, B-52H and FB-111A strike aircraft, augmented by 1,054 Titan II, Minuteman II and Minuteman III ICBMs.

In an effort to augment the Looking Glass mission, a 1973 initiative resulted in the establishment of the National Emergency Airborne Command Post (NEACP), also known as "knee cap," resulting in the procurement of four Boeing E-4 aircraft derived from the Boeing 747. The E-4 aircraft were originally stationed at Andrews AFB, Maryland so they could be easily accessed by the President and the Secretary of Defense, with SAC also establishing three dispersed support squadrons for the E-4 at Westover AFB, Massachusetts, Barksdale AFB, Louisiana and March AFB, California. Basing for the E-4 aircraft was later moved to Offutt AFB, with one E-4 continuously stationed at Andrews AFB in order to be available to the National Command Authority.

In 1985, after waiting decades for a new penetrating manned bomber, SAC took delivery of its first B-1B Lancer, while concurrent development of the black project that would eventually result in the Northrop Grumman B-2 Spirit also continued until that aircraft was officially unveiled to the public in late 1988. On 27 September 1991, as the Soviet Union was dissolving after the August 1991 Soviet coup d'état attempt, President George H. W. Bush began taking SAC bomber and associated aerial refuelling aircraft off continuous nuclear alert.[10]

SAC's final major operational engagement occurred during the 1990-1991 time frame during the First Gulf War. SAC bomber, tanker and reconnaissance aircraft flew numerous conventional bombing, aerial refueling and reconnaissance missions over and near Iraq from RAF Fairford and other bases in Great Britain, Turkey, Akrotiri, Cyprus, Diego Garcia, Saudi Arabia, and the United Arab Emirates.

Post–Cold War history

On 31 May 1992, following the collapse of the Soviet Union, and the end of the Cold War, SAC was eliminated in a major reorganization of USAF commands. Two of the air force's U.S.-based war-fighting commands, SAC and Tactical Air Command (TAC), were reorganized into a single organization, Air Combat Command (ACC). ACC was essentially given the combined missions that SAC and TAC held respectively, with the newly-designated Air Mobility Command (AMC) inheriting most of SAC's KC-135 Stratotanker and KC-10 Extender aerial refueling tanker force, while a small portion of KC-135 aircraft were reassigned to United States Air Forces in Europe (USAFE) and Pacific Air Forces (PACAF), the latter to include PACAF-gained KC-135 aircraft of the Alaska Air National Guard. SAC's former land-based ICBM force, initially part of ACC, eventually became part of the new Air Force Space Command (AFSPC). The USAF nuclear component was then officially combined with the United States Navy's strategic nuclear component, its Fleet Ballistic Missile (FBM) submarines, to form United States Strategic Command (USSTRATCOM), which is headquartered at SAC's former complex at Offutt AFB, Nebraska.

In late 2009, the ICBM force was transferred yet again, this time from AFSPC to the newly-established Air Force Global Strike Command (AFGSC). In early 2010, the B-2 Spirit and B-52 Stratofortress bomber force was also reassigned to AFGSC, while the B-1 Lancer bomber force remained in ACC due to the B-1's removal from the nuclear strike mission and reassignment to conventional roles only.[11]

The Strategic Air and Space Museum, formerly the SAC Museum, was located adjacent to Offutt AFB till moved to its site off of I-80 between Omaha and Lincoln, preserves SAC's heritage in a fashion open to public view.

Subordinate components

SAC included a large number of subordinate components. At the highest level, five Numbered Air Forces served within the command at various times, the Second Air Force, Eighth Air Force, Fifteenth Air Force, Sixteenth Air Force, and briefly, in 1991–92, the Twentieth Air Force. Large numbers of USAF Air Divisions served with the command and were typically respnsible for an average of three or four geographically separated wings. At lower levels, there were a large number of Strategic Air Command wings, groups such as the 1st Combat Evaluation Group, and large numbers of bases (see List of Strategic Air Command Bases).

Strategic Air Command in the United Kingdom was among the command's largest overseas concentrations of forces, with additional forces at bases in North Africa during the 1950s and 1960s in addition to SAC bomber, tanker, and/or reconnaissance aircraft assets at the former Ramey AFB, Puerto Rico in the 1950s, 1960s and 1970s, and at Andersen AFB, Guam, RAF Mildenhall, United Kingdom and the former NAS Keflavik, Iceland through the 1990s. SAC "Provisional" wings were also located in Okinawa and Thailand during the Vietnam War and at Diego Garcia and in the United Kingdom during the first Gulf War.

Myriad smaller subunits included test, evaluation and acquisition activities serving as tenants with former Air Force Systems Command and Air Force Logistics Command entities, as well as ceremonial guard formations such as the SAC Elite Guard.

Wings of the command included:[12]

- List of ANG wings assigned to Strategic Air Command
- List of USAF Bomb Wings and Wings assigned to Strategic Air Command
- List of USAF Fighter Wings assigned to Strategic Air Command
- List of USAF Provisional Wings assigned to Strategic Air Command
- List of USAF reconnaissance wings assigned to Strategic Air Command
- List of USAF Strategic Wings assigned to the Strategic Air Command

Strategic Air Command insignia

The insignia of SAC was designed in 1951 by Staff Sergeant R.T. Barnes, then assigned to the 92nd Bombardment Wing. Submitted in a command-wide contest, it was chosen as the winner by a three judge panel: General Curtis E. LeMay, Commander-in-Chief, Strategic Air Command [CINCSAC]; General Thomas S. Power, Vice Commander-in-Chief, Strategic Air Command; and Brigadier General AW Kissner, Chief of Staff, Strategic Air Command. Staff Sergeant Barnes' winning design netted him a $100 United States Savings Bond.[13] [14]

It has a sky-blue field with two white shaded blue-gray clouds, one in the upper left and one in the lower right extending to the edges of the shield. Upon this is a cubit arm in armor issuing from the lower right and extending toward the upper left part of the shield. The hand is grasping a green olive branch, and three red lightning bolts.

The blue sky is representative of USAF operations. The arm and armor are a symbol of strength, power and loyalty and represents the science and art of employing far-reaching advantages in securing the objectives of war. The olive branch, a symbol of peace, and the lightning flashes, symbolic of speed and power are qualities underlying the mission of the Strategic Air Command.[15]

The blue background of the SAC crest meant that SAC's reach was through the sky and that it was global in scope. The clouds meant that SAC was all-weather capable. The mailed fist depicted force, symbolized by lightning bolts of destruction. The olive branch represents peace.

In addition to the SAC crest, non-camouflaged SAC aircraft bore the SAC Stripe. The stripe consisted of a very dark blue background speckled with stars. The stripe appeared on the sides of SAC aircraft in the area of the cockpit on bomber aircraft and mid-fuselage on tanker and command post aircraft running from the top to the bottom of the fuselage at an angle from 11 o'clock to 5 o'clock. The stripe also appeared on ICBMs in the strategic missile force. The SAC crest was a bit wider than the stripe and was placed over the stripe. The stripe indicated that SAC was always ready to fulfill its mission.

Aircraft and missiles

Aircraft — primary mission

Boeing B-52D, AF Serial No. 56-0687 on display at B-52 Memorial Park, Orlando International Airport, Florida (formerly McCoy Air Force Base, Florida). Photo taken April 4, 2003.

- B-1 Lancer from 1986–92; includes Air National Guard (ANG) from 1989–92
- B-17 Flying Fortress (RB-17G) from 1946–51
- B-26 Invader (RB-26) from 1949–50
- B-29 Superfortress from 1946–53
- B-36 Peacemaker from 1948–58
- B-45 Tornado from 1950–53
- B-47 Stratojet 1951–65
- B-50 Superfortress from 1948–54
- B-52 Stratofortress from 1955–92; includes Air Force Reserve (AFRES) from 1989–92
- B-57 Canberra from 1956–62
- B-58 Hustler from 1960–69
- C-119 Flying Boxcar from 1956–73
- DC-130 Hercules from 1966–76
- E-4 Nightwatch from 1975–92
- EC-135 Looking Glass from 1963–92
- F-2 Expeditor (F=Fotorecon; C-45)
- F-6 Mustang (F=Fotorecon; P-51)
- F-9 Flying Fortress (F=Fotorecon; B-17F/G)
- F-13 Superfortress (F=Fotorecon; B-29A)
- F-47 Thunderbolt (P-47) from 1946–47
- F-51 Mustang (P-51) from 1946–49
- F-82 Twin Mustang from 1947–50
- F-80 Shooting Star from 1946–48
- F-84 Thunderjet 1948–57
- F-86 Sabre 1949–50
- F-102 Delta Dagger 1960
- FB-111 Aardvark from 1969–90
- KC-10 Extender from 1981–92
- KB-29 from 1949–56
- KC-97 Stratotanker SAC from 1951–64 and SAC-gained ANG from 1973–77
- KC-135 Stratotanker SAC from 1957–91; includes ANG and AFRES from 1975–92
- RC-45 Expeditor
- RC-135 Rivet Joint/Rivet Brass/Rivet Amber/Rivet Card/Rivet Ball/Cobra Ball/Cobra Eye/Combat Sent
- SR-71 Blackbird from 1966–91
- U-2 Dragon Lady from 1962–91
- TR-1 Dragon Lady from 1989–91
- UC-45 Expeditor

Aircraft — support

- AT-11 Kansan
- B-26 Invader from 1949–50
- C-45 Expeditor from 1946–51
- C-47 Skytrain from 1946–47
- C-54 Skymaster from 1946–75
- C-82 Packet from 1946 through 1947
- C-97 Stratofreighter from 1949–78
- C-118 Liftmaster from 1957–75
- C-124 Globemaster II from 1959–62
- C-131 Samaritan
- C-135 Stratolifter
- CH-3 Sea King SAC c. 1960s
- HU-16 Albatross SAC c. 1950s/1960s
- L-4 Grasshopper from 1949–50
- L-5 Sentinel from 1949–50
- L-13 Grasshopper from 1949–50
- PBY Catalina (AAF designation OA-10 Catalina) from 1946–47
- T-38 Talon from 1981–91
- UH-1 Huey from 1966–92

Missiles fielded by Strategic Air Command

- ADM-20 Quail
- AGM-28 Hound Dog
- AGM-69 SRAM
- AGM-84 Harpoon
- AGM-86 Air Launched Cruise Missile
- AGM-129 Advanced Cruise Missile
- HGM-16 Atlas
- HGM-25A Titan I
- LGM-25 Titan II
- LGM-30A/B Minuteman I
- LGM-30F Minuteman II
- LGM-30G Minuteman III
- LGM-118A Peacekeeper
- SM-62 Snark
- PGM-17A Thor
- PGM-19A Jupiter

Titan II missile launching from silo.

Notes

[1] "AIR FORCE GLOBAL STRIKE COMMAND (USAF)" (http://www.afhra.af.mil/factsheets/factsheet.asp?id=15047). Air Force Historical Research Agency. 17 July, 2009. . Retrieved 17 January 2012.
[2] Boyne, Walter J (1997), *Beyond The Wild Blue: A History of the United States Air Force 1947–1997*, New York: St. Martin's Press.
[3] See Walton S. Moody, 'Building a Strategic Air Force,' Air Force History and Museums Program, 1995, pp 60, 62. The 40th and 444th Groups were also earmarked for assignment to the 58th Wing. See also 'Air Force Combat Units of World War II' and 'Air Force Combat Wings: Lineage and Honors Histories, 1947-1977,' USAF books accessible via the Air Force Historical Studies Office index of titles (http://newpreview.afnews.af.mil/afhso/booksandpublications/titleindex.asp).
[4] Walton S. Moody, 'Building a Strategic Air Force,' 1995, 78, drawing on Ltr, Maj Gen St. C. Streett, Dep CG SAC, to CG AAF, subj: Operational Training and Strategic Employment of Strategic Air Command, Jul 25, 1946, SAC/HO.
[5] Tillman, Barrett (2007), *LeMay*, New York: Palgrave Macmillan, p. 94.
[6] See also Robert J Boyd, 'SAC's fighter planes and their operations', Offutt AFB, NE : Office of the Historian, Headquarters Strategic Air Command ; Washington, D.C. : Supt. of Docs, U.S. G.P.O., [1988]
[7] Rosenberg, David A (June 1979), "American Atomic Strategy and the Hydrogen Bomb Decision" (http://www.jstor.org/stable/1894674), *The Journal of American History*, pp. 62–87, , retrieved 1 March 2009, 65.
[8] Rosenberg, David A (Spring 1983), "The Origins of Overkill: Nuclear Weapons and American Strategy, 1945–1960" (http://www.jstor.org/stable/2626731), *International Security*, Los Angeles: University of Southern California, pp. 3–71, , retrieved 1 March 2009, 17.
[9] Pike, John (24 July 2011). "Strategic Air Command" (http://www.globalsecurity.org/wmd/agency/sac.htm). Global Security. . Retrieved 28 November 2011.
[10] http://www.stratofortress.org/history.htm
[11] "Factsheet" (http://www.af.mil/information/factsheets/factsheet.asp?id=16613), *Information*, USA: Air Force, .
[12] See Mixer, Ronald E., Genealogy of the STRATEGIC AIR COMMAND, Battermix Publishing Company, 1999, and Mixer, Ronald E., STRATEGIC AIR COMMAND, An Organizational History, Battermix Publishing Company, 2006
[13] "History" (http://www.7bwb-36assn.org/b36genhistpg3.html), *7BWB-36ASSN*, .
[14] "SAC" (http://b-29s-over-korea.com/SAC/sac2.html), *B-29s over Korea*, .
[15] "Shield" (http://www.strategic-air-command.com/patches/cmd-SAC-0-shield.htm), *Strategic Air Command*, .

References

- Boyne, Walter, Beyond The Wild Blue: A History of the United States Air Force 1947–1997, New York: St. Martin's Press, 1997.
- Moody, Walton S. Dr., *Building a Strategic Air Force*, U.S. Government Printing Office, 1998.
- Rosenberg, David A (June 1979), "American Atomic Strategy and the Hydrogen Bomb Decision", The Journal of American History, pp. 62–87
- Tillman, Barrett, *LeMay,* New York: Palgrave Macmillan, 2007.

Further reading

- Adams, Chris, *Inside The Cold War; A Cold Warrior's Reflections*, Air University Press, 1999; 2nd printing 2004; 3rd printing 2005.
- Adams, Chris, "Ideologies in Conflict; A Cold War Docu-Story,Writers' Showcase, New York, 2001.
- Clark, Rita F. Major, *From Snark to Peacekeeper*, Office of the Historian, HQ. SAC, Offutt AFB. NE. 1990.
- Clark, Rita F. Major, *SAC Missile Chronology 1939–1988*, Office of the Historian, HQ. SAC, Offutt AFB. NE. 1988.
- Clark, Rita F. Major, *Strategic Air Command*, U.S. Government Printing Office.
- Goldberg, Sheldon A., *The Development of the Strategic Air Command*, Office of the Historian, HQ. SAC, Offutt AFB. NE. 1986.
- Knaack, Marcelle Size, *Post-World War II Bombers 1945–1973*, Office of Air Force History, United States Air Force, Washington DC 1988.
- Knaack, Marcelle Size, *Post-World War II Fighters 1945–1973*, Office of Air Force History, United States Air Force, Washington DC 1986.
- Lloyd, Alwyn T., *B-47 Stratojet in detail & scale*, TAB Books, 1988.

- Lloyd, Alwyn T. *A Cold War Legacy: A Tribute to Strategic Air Command, 1946-1992*. Missoula, Mont: Pictorial Histories Pub, 2000. ISBN 1-57510-052-5.
- Mixer, Ronald E., *Genealogy of the Strategic Air Command*, Battermix Publishing Company, 1999
- Mixer, Ronald E., *Strategic Air Command, An Organizational History*, Battermix Publishing Company, 2006.
- Narducci, Henry M (1 April 1988). Strategic Air Command and the Alert Program: A Brief History (http://www.siloworld.net/DOWNLOADS/SAC Brief History Reduced.pdf) (Report). Offutt Air Force Base: Office of the Historian, Headquarters Strategic Air Command. Retrieved 18 October 2011.
- Polmar, Norman, *Strategic Air Command*, 1st Edition, Nautical & Aviation Publishing, 1954
- Polmar, Norman, *Strategic Air Command*, 2nd Edition, Nautical & Aviation Publishing, 1996.
- Ravenstein, Charles, A., *Air Force Combat Wings 1947–1977*, Office of Air Force History, USAF, 1984.
- Yenne, Bill, *History of the U.S. Air Force*, Exeter Books, 1990.
- Yenne, Bill, SAC, A Primer of Modern Strategic Airpower, Presido Press, 1992.

External links

- Interview with SAC Lt. General James Edmundson (transcripts from the PBS program *Race for the Superbomb* (http://www.pbs.org/wgbh/amex/bomb/)):
  - The Mission of the Strategic Air Command (http://www.pbs.org/wgbh/amex/bomb/filmmore/reference/interview/edmund06.html)
  - Strategic Air Command Bomber Response Time (http://www.pbs.org/wgbh/amex/bomb/filmmore/reference/interview/edmund05.html)
  - Strategic Air Command's War Plans (http://www.pbs.org/wgbh/amex/bomb/filmmore/reference/interview/edmund11.html)
  - Strategic Air Command's Role in Ending the Cold War (http://www.pbs.org/wgbh/amex/bomb/filmmore/reference/interview/edmund13.html)
- Air Force History Part Three: Countering the Communist Threat During the Cold War (http://www.airforce.com/learn-about/history/part3/)
- Strategic-Air-Command.com (http://www.strategic-air-command.com/).
- Strategic Air Command Memorial Amateur Radio Club (http://www.sacmarc.org)
- SAC Elite Guard Association (http://www.saceliteguard.com)

Media

- The short film *15 AF HERITAGE - HIGH STRATEGY - BOMBER AND TANKERS TEAM (1980)* (http://www.archive.org/details/gov.dod.dimoc.52240) is available for free download at the Internet Archive [*more*]
- The 1958 *Air Force Special Film Project 416, "Power of Decision"* (http://www.archive.org/details/AirForceSpecialFilmProject416powerOfDecision) is available at the Internet Archive
- The short film *STRATEGIC AIR COMMAND SEMIANNUAL FILM REPORT (1968)* (http://www.archive.org/details/gov.dod.dimoc.20791) is available for free download at the Internet Archive [*more*]
- The short film *Air Force Special Film Project 1236, "SAC Command Post"* (http://www.archive.org/details/AirForceSpecialFilmProject1236sacComnandPost) is available for free download at the Internet Archive [*more*]
- The short film *THE STRENGTH OF SAC (1966)* (http://www.archive.org/details/gov.dod.dimoc.26474) is available for free download at the Internet Archive [*more*]
- The short film *SAC - THE GLOBAL SHIELD (1980)* (http://www.archive.org/details/gov.dod.dimoc.51341) is available for free download at the Internet Archive [*more*]
- The short film *AIR FORCE STORY, VOL 2, CHAPTER 5 -- OUR WORLDWIDE AIR FORCE, 1953-1959 (1959)* (http://www.archive.org/details/gov.dod.dimoc.38551) is available for free download at the Internet Archive [*more*]

United_States_Strategic_Command

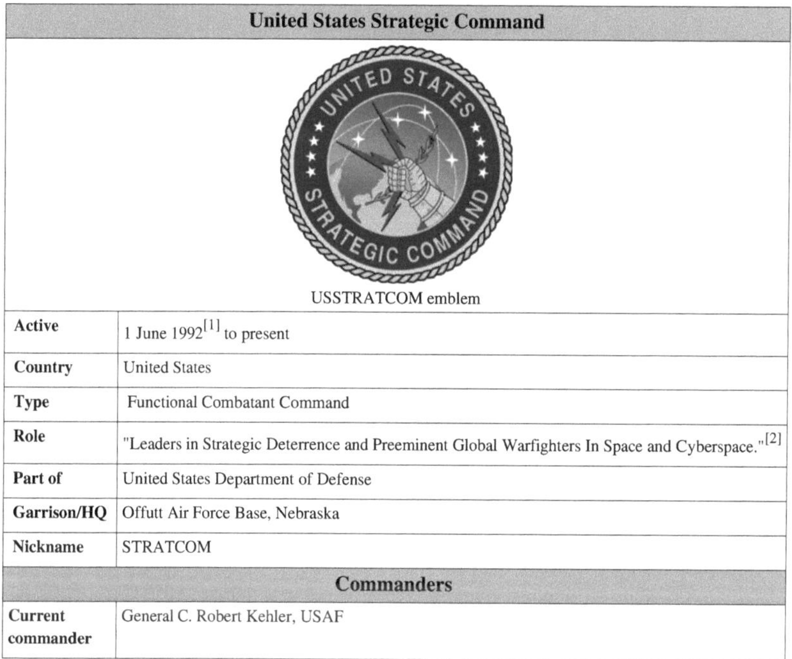

United States Strategic Command	
	USSTRATCOM emblem
Active	1 June 1992[1] to present
Country	United States
Type	Functional Combatant Command
Role	"Leaders in Strategic Deterrence and Preeminent Global Warfighters In Space and Cyberspace."[2]
Part of	United States Department of Defense
Garrison/HQ	Offutt Air Force Base, Nebraska
Nickname	STRATCOM
Commanders	
Current commander	General C. Robert Kehler, USAF

United States Strategic Command (**USSTRATCOM**) is one of nine Unified Combatant Commands of the United States Department of Defense (DoD). It is charged with space operations (such as military satellites), information operations (such as information warfare), missile defense, global command and control, intelligence, surveillance, and reconnaissance (C^4ISR), global strike and strategic deterrence (the United States nuclear arsenal), and combating weapons of mass destruction.

Strategic Command was established in 1992 as a successor to Strategic Air Command (SAC). It is headquartered at Offutt Air Force Base south of Omaha, Nebraska. In October 2002, it merged with the United States Space Command (USSPACECOM). It employs more than 2,700 people, representing all four services, including DoD civilians and contractors.

Strategic Command is one of the three Unified Combatant Commands organized along a functional basis. The other six are organized on a geographical basis. The unified military combat command structure is intended to give the President and the Secretary of Defense a unified resource for greater understanding of specific threats around the world and the means to respond to those threats as quickly as possible.

History

On 1 June 1992, President George H. W. Bush established the U.S. Strategic Command from the Strategic Air Command (SAC) and other Cold War military bodies, now obsolete due to the change in world politics. The Command unified planning, targeting and wartime employment of strategic forces under one commander. Day-to-day training, equipment and maintenance responsibilities for its forces remained with the Air Force and Navy.

As a result of the 2002 Nuclear Posture Review, the Cold War system of relying solely on offensive nuclear response was modified. Shortly after a meeting between President George W. Bush and Russian President Vladimir Putin in Moscow in May 2002, a summit was held during which both leaders signed a treaty promising bilateral reductions that would result in a total of 1,700 to 2,200 operationally deployed strategic nuclear weapons for each country by the year 2012.

Space and Global Strike reorganization

The activation of the new USSTRATCOM took place on 1 October 2002. The merged command was responsible for both early warning of and defense against missile attack as well as long-range strategic attacks.

President George W. Bush signed Change Two to the Unified Command Plan on 10 January 2003, and tasked USSTRATCOM with four previously unassigned responsibilities: global strike, missile defense integration, Department of Defense Information Operations, and C^4ISR (command and control, communications, computers, intelligence, surveillance and reconnaissance). This combination of roles, capabilities and authorities under a single unified command was unique in the history of unified commands.

After some consideration concerning the separation of the JFCC for Space and Global Strike missions, according to AirForceTimes.com[3] and InsideDefense.com,[4] In 2005, General Cartwright began the process of separating the JFCC for Space and Global Strike into two individual JFCCs: a JFCC for Space (JFCC Space) and a JFCC for Global Strike and Integration (JFCC GSI).[5] U.S. Strategic Command officials were expected to deliver a detailed plan on the separation to General Cartwright for approval by September 2006.[6]

Some officials believed this would allow each to focus more effectively on its primary mission and allow the mission of space to have focused attention and be better integrated with other military capabilities. This comes after some concern by officials and lawmakers such as U.S. Senator Wayne Allard (R-Colo.), an advocate for national security space activities, complained in a March 2006 memo to Defense Secretary Donald Rumsfeld about what he saw as a declining emphasis on space within the U.S. Department of Defense and specifically the way space has been organized at U.S. Strategic Command.[7]

As result of the separation, The Missile Correlation Center in Cheyenne Mountain AFS was broken into two separate entities. NORAD/NORTHCOM (N2C2) now controls the Missile and Space Domain (MSD) and JFCC Space controls the Missile Warning Center (MWC). They are both still located at Cheyenne Mountain AFS. It was expected that MSD would eventually move to Peterson AFB to join the rest of N2C2.

Mission statement

The LeMay building

USSTRATCOM promotes global security for America: The missions of U.S. Strategic Command are to deter attacks on U.S. vital interests, to ensure U.S. freedom of action in space and cyberspace, to deliver integrated kinetic and non-kinetic effects to include nuclear and information operations in support of U.S. Joint Force Commander operations, to synchronize global missile defense plans and operations, to synchronize regional combating of weapons of mass destruction plans, to provide integrated surveillance and reconnaissance allocation recommendations to the SECDEF, and to advocate for capabilities as assigned.

Subordinate Commands

Army

- United States Army Space and Missile Defense Command (SMDC)

Air Force

- Air Force Global Strike Command (AFGSC)
- Air Force Space Command (AFSC)

Primary operational units

USSTRATCOM exercises command authority over four joint functional component commands, also known as JFCCs as well as Joint Task Forces and Service Components. This combination of authorities, oversight, leadership and management is supposed to enable a more responsive, flattened organizational construct according to the commands leadership.

- **Joint Functional Component Commands** These commands are responsible for the day-to-day planning and execution of primary mission areas: space and global strike; intelligence, surveillance and reconnaissance; network warfare; integrated missile defense; and the recently added mission of combating weapons of mass destruction.
  - Joint Functional Component Command for Global Strike (JFCC-GS) The Commander Eighth Air Force (AFSTRAT-GS) serves as the Joint Functional Component Commander for Global Strike. JFCC-GS conducts planning, integration, execution and force management of assigned missions of deterring attacks against the U.S., its territories, possessions and bases, and should deterrence fail, by employing appropriate forces. Some of these tasks belonged to a JFCC for Space and Global Strike before being split into two components.
  - Joint Functional Component Command for Space (JFCC Space) The Commander 14th Air Force (AFSTRAT-SP) serves as the commander for JFCC Space. This component conducts planning, execution, and force management, as directed by the commander of USSTRATCOM, of the assigned missions of coordinating, planning, and conducting space operations.
  - Joint Functional Component Command for Integrated Missile Defense (JFCC IMD)—The Commander, U.S. Army Space and Missile Defense Command/Army Forces Strategic Command, also serves as the commander for the JFCC IMD. This component is responsible for meeting USSTRATCOM's Unified Command Plan responsibilities for planning, integrating, and coordinating global missile defense operations and support. JFCC IMD conducts the day-to-day operations of assigned forces and coordinates activities with associated

combatant commands, other USSTRATCOM Joint Functional Components and the efforts of the Missile Defense Agency.

- Joint Functional Component Command for Intelligence, Surveillance and Reconnaissance (JFCC-ISR)—The Commander, JFCC-ISR, also serves as the Director, Defense Intelligence Agency. This component is responsible for coordinating global intelligence collection to address DoD worldwide operations and national intelligence requirements. It will serve as the center for planning, execution and assessment of the military's global Intelligence, Surveillance, and Reconnaissance operations; a key enabler to achieving global situational awareness.
- Center for Combating Weapons of Mass Destruction (SCC WMD)—The Secretary of Defense recently assigned USSTRATCOM responsibility for integrating and synchronizing DoD's efforts for combating weapons of mass destruction. SCC WMD works closely with the Defense Threat Reduction Agency (DTRA) and declared Initial Operating Capability on 26 January 2006 in a ceremony in Washington, D.C.[8]
- United States Cyber Command (USCYBERCOM)—The CYBERCOM is a subordinate unified command under United States Strategic Command created by United States Defense Secretary Robert Gates on 23 June 2009, and activated in September of that year. The command is led by the director of the National Security Agency, General Keith B. Alexander. It combined JTF-GNO and JFCC-NW, which were dissolved in October 2010.

Task forces

USSTRATCOM relies on various task forces for the execution of its global missions. These include:

- Aerial Refueling/Tankers—Task Force 294—Air Force refueling aircraft greatly enhance the command's capability to conduct global combat and reconnaissance operations. Tankers are assigned to Eighteenth Air Force, Scott AFB, Illinois, with headquarters at Air Mobility Command, Scott AFB, Illinois.
- Airborne Communications—The Navy's E-6B Mercury aircraft provide a survivable communications link between national decision-makers and the nation's strategic forces. An airborne command post, the E-6B enables the President and the Secretary of Defense to directly contact crews on the nation's ballistic missile submarines, land-based intercontinental ballistic missiles and long-range bombers. E-6B aircraft are assigned to Strategic Communications Wing One (TACAMO), Tinker AFB, Oklahoma.
- Ballistic Missile Submarines—Considered the most survivable leg of the nation's strategic forces, Navy ballistic missile submarines, or SSBNs, provide launch capability from around the globe using the Trident missile weapon system. Atlantic SSBNs are based at Kings Bay Submarine Base, Georgia, with headquarters at Commander, Submarine Forces U.S. Atlantic Fleet, Naval Base Norfolk, Virginia; Pacific SSBNs are based at Bangor, Washington, with headquarters at Commander Submarine Forces U.S. Pacific Fleet, Pearl Harbor Naval Base, Hawaii. Task Forces 134 and 144 are operationally assigned, with 134 being for Pacific, 144 for Atlantic.
- Joint Task Force-Global Network Operations (JTF-GNO)—Located in Arlington, Va., the Joint Task Force-Global Network Operations (JTF-GNO) is U.S. Strategic Command's operational component engaged in operation and limited defense of the DoD's Global Information Grid— supporting JFCC-NW in fighting Cyber-terrorism directed against the US military. This is done by integrating GNO capabilities into the operations of all DoD computers, networks, and systems used by DoD combatant commands, services and agencies. . The Director, Defense Information Systems Agency also heads the Joint Task Force for Global Network Operations. This organization is responsible for operating and defending U.S. worldwide information networks, a function closely aligned with the efforts of the Joint Functional Component Command for Network Warfare, commanded by Director, National Security Agency.
- Strategic Bomber and Reconnaissance Aircraft—Aircraft assigned to Eighth Air Force, Barksdale AFB, Louisiana, are capable of deploying air power to any area of the world. B-1B Lancer heavy bombers are available at Dyess AFB, Texas and Ellsworth AFB, South Dakota, though the United States does not carry nuclear weapons in the B-1B in compliance with international treaty. B-52 Stratofortress heavy bombers are based at Barksdale

AFB, Louisiana, and Minot AFB, North Dakota. B-2 Spirit stealth bombers are stationed at Whiteman AFB, Missouri. Worldwide reconnaissance aircraft assigned to Eighth Air Force that support the USSTRATCOM mission include the RC-135 Rivet Joint, Offutt AFB, Nebraska, and the U-2S Dragon Lady, Beale AFB, California.

- Land-based Intercontinental Ballistic Missiles—Air Force ICBMs dispersed in hardened silos across the nation's central tier, provide a quick-reacting and highly reliable component to the nation's strategic forces. Minuteman III missile launch control centers are based from F.E. Warren AFB, Wyoming; Malmstrom AFB, Montana; and Minot AFB, North Dakota, Peacekeeper missiles were based at F.E. Warren AFB. ICBM crews report to Twentieth Air Force, F.E. Warren AFB which is also dual hatted as Task Force 214, which reports to U.S. Strategic Command. The Peacekeeper missiles were officially deactivated on 19 September 2005.[9] Targeting and strategic communications are provided by the 625th Strategic Operations Squadron (625 STOS).

Leadership

In 2007, General Kevin P. Chilton took over command of USSTRATCOM. He served as the senior commander of the joint military forces from the four branches of the military assigned to the command. He is the leader, steward and advocate of the nation's strategic capabilities.

His responsibilities include integrating and coordinating the necessary command and control capability to provide support with the most accurate and timely information for the President of the United States, the Secretary of Defense, and to regional combatant commanders.

On 7 May 2009, Chilton stated that the United States would not be constrained in its response to a cyber attack, therefore demonstrating the utility of his command which combines cyber defense with global strike.[10]

List of Combatant Commanders

No.	Image	Name	Start of Term	End of Term
1.		General George L. Butler, USAF	1992	1994
2.		Admiral Henry G. Chiles, Jr., USN	1994	1996
3.		General Eugene E. Habiger, USAF	1996	1998
4.		Admiral Richard W. Mies, USN	1998	2002
5.		Admiral James O. Ellis, Jr., USN	2002	2004
6.		General James E. Cartwright, USMC	2004	2007

Acting		Lt. Gen C. Robert Kehler, USAF	4 August 2007	17 October 2007
7.		General Kevin P. Chilton, USAF	2007	2011
8.		General C. Robert Kehler, USAF	2011	*Present*

Innovations

A previous commander, General James Cartwright (2004–07), explored ways to incorporate innovative collaborative tools into what has traditionally been considered a very centralized military organization. Speaking at a convention Cartwright said, "Where I would like to be is well outside the comfort zone of my organization. But what we've started with is just some simple 'blogging' tools, to try to change the culture a little bit; to try to allow people to contribute."

See also

- Nuclear weapons and the United States

References

[1] Air Force Magazine (http://www.airforce-magazine.com), Journal of the Air Force Assoc.,**Space Almanac 2008**, August 2008

[2] Stratcom Homepage (http://www.stratcom.mil/mission/)

[3] "?" (http://www.airforcetimes.com/story.php?f=1-292925-1995790.php). .

[4] "?" (http://defense.iwpnewsstand.com/cs_newsletters.asp?NLN=MISSILE&ACTION=RECENT). .

[5] "Air Force Times" (http://www.airforcetimes.com/story.php?f=1-292925-1995790.php). Air Force Times. . Retrieved 19 May 2011.

[6] Inside Defense NewsStand: *Inside Missile Defense* 21 November 2007, Vol. 13, No. 24 (http://defense.iwpnewsstand.com/cs_newsletters.asp?NLN=MISSILE&ACTION=RECENT)

[7] Space News: *U.S. Sen. Wayne Allard (R-Colo.) on US Dept of Defense*, 29 November 2007 (http://www.space.com/spacenews/archive06/Editorial_031306.html)

[8] "National News" (http://www.dcmilitary.com/navy/seaservices/10_54/national_news/39450-1.html). .

[9] Billings Gazette: Wyoming Peacekeeper Missile System (http://www.billingsgazette.com/index.php?id=1&display=rednews/2005/09/19/build/wyoming/50-missle-system.inc)

[10] "Cyber attack could bring U.S. military response" (http://www.securityfocus.com/brief/961/). Securityfocus.com. 11 May 2009. . Retrieved 19 May 2011.

- The Unified Command Plan (http://www.dtic.mil/doctrine/jel/jfq_pubs/1231.pdf), Joint Forces Quarterly; Johnson, Spencer (September 2002).
- United States Strategic Command Official Website (http://www.stratcom.mil/)
- Air Force Magazine (http://www.airforce-magazine.com), Journal of the Air Force Assoc., August 2008.

External links

- FAS: United States Space Command (USSPACECOM) (http://www.fas.org/spp/military/program/nssrm/initiatives/usspace.htm)
- GAO Report: Additional Actions Needed by U.S. Strategic Command to Strengthen Implementation of Its Many Missions and New Organization (http://www.gao.gov/new.items/d06847.pdf)

Retroactive_continuity

Retroactive continuity (**retcon** for short)[1] is the alteration of previously established facts in a fictional work.[2] Retcons are done for many reasons, including the accommodation of sequels or further derivative works in a series, wherein newer authors or creators want to revise the in-story history to allow a course of events that would not have been possible in the story's original continuity. Other reasons might be the reintroduction of popular characters, resolution of errors in chronology, the updating of a familiar series for modern audiences, or simplification of an excessively complex continuity structure.

Retcons are common in pulp fiction, especially comic books published by long-established houses such as DC, Marvel and leading manga publishers. The long history of popular titles and the plurality of writers who contribute stories can often create situations that demand clarification or revision of exposition. Retcons also appear in soap operas, serial drama, movie sequels, professional wrestling, video games, radio series, and other kinds of serial fiction. Retcons have been criticized as "cheating" on the part of the author, seen as an effort to purge "unpopular" elements from the storyline and force literary fads upon the audience, thus hurting suspension of disbelief.

Origins of the term

The first published use of the phrase "retroactive continuity" is found in Elgin Frank Tupper's 1974 book *The theology of Wolfhart Pannenberg*.[3]

> Pannenberg's conception of **retroactive continuity** ultimately means that history flows fundamentally from the future into the past, that the future is not basically a product of the past.

The first known printed use of "retroactive continuity" as referring to the altering of history within a fictional work is in *All-Star Squadron* #18 (cover-dated February 1983) from DC Comics. The series was set on DC's Earth-Two, an alternative universe in which Golden Age comic characters proceed and age subsequent to their first appearances in real time. Thus by the early 1980s Superman was in his 60s and the Batman had died and been succeeded by his daughter, The Huntress, whereas the Superman and Batman of Earth-One, DC's primary universe, are perpetually young to early middle-age adults. *All-Star Squadron* in particular, was set during World War II on Earth-Two, so it was in the past of an alternative universe, thus all its events had repercussions on the contemporary continuity of the DC multiverse. Each issue literally changed the history of the fictional world in which it was set. In the letters column, a reader remarked that the comic "must make you [the creators] feel at times as if you're painting yourself into a corner," and "Your matching of Golden Age comics history with new plotlines has been an artistic (and I hope financial!) success."

Writer Roy Thomas responded, "we like to think that an enthusiastic ALL-STAR booster at one of Adam Malin's Creation Conventions in San Diego came up with the best name for it a few months back: 'Retroactive Continuity'. Has kind of a ring to it, don't you think?"[4] The term, possibly in limited similar use before *All-Star Squadron* #18, then took firm root in the consciousness of fans of American superhero comics.

"Retroactive continuity" was shortened to "retcon", reportedly by Damian Cugley in 1988 on USENET. Hard evidence of Cugley's abbreviation has yet to surface, though in a USENET posting on August 18, 1990, Cugley posted a reply in which he identified himself as "The originator of the word 'retcon'."[5] Cugley used the newly-shortened word to describe a development in the comic book *Saga of the Swamp Thing*, which reinterprets the events of the title character's origin by revealing facts that, up to that point, are not part of the narrative and were not intended by earlier writers. In this case, the revelation is that the titular character's memories are false and he is not who he thinks he is.[6] Alan Moore's retcons often involve false memories, for example *Marvelman* (aka *Miracleman* in America), and *Batman: The Killing Joke*.

Types

Although there is considerable ambiguity and overlap between different kinds of retcon, there are some distinctions that fans have made between them, depending on whether the retcon in question adds to, alters, removes, or restarts' material from the narrative's continuity. These distinctions often evoke different reactions from fans of the material.

Addition

Some retcons do not directly contradict previously established facts, but "fill in" missing background details, usually to support current plot points. This was the sense in which Thomas used "retroactive continuity", as a purely additive process that did not "undo" any previous work, a common theme in his work on *All-Star Squadron*. Kurt Busiek took a similar approach with *Untold Tales of Spider-Man*, a series which told stories that specifically fit between issues of the original *The Amazing Spider-Man* series, sometimes explaining discontinuities between those earlier stories. John Byrne utilized a similar structure as well with *X-Men: The Hidden Years*.

Related to this is the concept of shadow history or secret history, in which the events of a story occur within the bounds of already-established events (especially real-world historical events), revealing a different interpretation of (or motivation for) the events. Some of Tim Powers novels are examples of this, such as *Last Call*, which suggests that Bugsy Siegel's actions were due to his being a modern-day Fisher King.

Alan Moore's additional information about the Swamp Thing's origins did not contradict or change any of the events depicted in the character's previous appearances, but changed the reader's interpretation of them. Such additions and reinterpretations are very common in *Doctor Who*.[2]

In the *Star Trek* franchise, the books *The Rise and Fall of Khan Noonien Singh (Volumes 1 & 2)*, by Greg Cox, detail the fictional Eugenics Wars of the early 1990s – still many years into the future when first mentioned in the episode "Space Seed" in 1967 – giving alternative explanations for real world events such as the Indian nuclear test of 1974 and the violent breakup of Yugoslavia in the early 1990s, presenting them as small parts of a single wider conflict.

The History Monks appear in Terry Pratchett's *Thief of Time* and explain anachronisms in the Discworld (such as Elizabethan theatre existing simultaneously with opera) by describing how history was previously destroyed by a magical clock and they have been haphazardly attempting to reconstruct it.

In the *Stargate* franchise, the TV show that followed the original film established the main antagonist from that film, the now-deceased Ra, as part of an alien species called the Goa'uld, establishing many more instances of that species as enemies as the series progressed.

Alteration

Retcons often add information that effectively states "what you saw isn't what really happened" and then introduce a different version of the backstory. This is usually interpreted by the audience as an overt change rather than a mere addition. The most common form this takes is when a character shown to have died (sometimes explicitly) is later revealed to have survived somehow. This is well known in horror films, which may end with the death of the monster, but when the film becomes successful, the studio plans a sequel, revealing that the monster survived after all. The technique is common in superhero comics,[2] where it has been used so frequently that the term *comic book death* has been coined for it.

An early famous example in popular culture is the return of Sherlock Holmes: writer Arthur Conan Doyle killed off the popular character in an encounter with his foe Professor Moriarty, only to bring Holmes back, due in large part to audience response.[7]

J. R. R. Tolkien in *The Hobbit* described the circumstances in which Bilbo Baggins won a magic ring from Gollum. However, by the time he wrote the sequel, *The Lord of the Rings*, his concept of the ring's nature had changed, at odds with the previous depiction. To explain this discrepancy, Tolkien retold this incident in the new work, explaining the original version as a lie inspired by the malevolent influence of the ring.

In many of his detective novels, Rex Stout implies that his character Nero Wolfe was born in Montenegro, and gives some details of his early life in the Balkans prior to and during World War I. However, in *Over My Dead Body* (1939), Wolfe tells an FBI agent that he was born in the United States. Stout revealed the reason for the change in a letter obtained by his authorized biographer, John McAleer: "In the original draft of *Over My Dead Body* Nero was a Montenegrin by birth, and it all fitted previous hints as to his background; but violent protests from *The American Magazine*, supported by Farrar & Rinehart, caused his cradle to be transported five thousand miles."[8]

In the 1950 novel *Pebble in the Sky*, science fiction novel writer Isaac Asimov depicted a future Earth with a largely radioactive crust, humans precariously surviving in the uncontaminated areas in between. There was the clear implication (though it was not explicitly stated) that this was the result of a nuclear war hundreds or thousands of years before the time of the plot. It was, however, pointed out by critics that such an extensive use of nuclear weapons as to leave persistent and widespread radiation even after centuries would have completely destroyed all life on Earth at the moment when it took place. Therefore, in a much later book, *Robots and Empire* (1985), Asimov provided a different origin for the future Earth's radioactivity - making it the result of a gradual process in the course of which human and other life could survive.

Fans may invent unofficial explanations for inconsistencies, the challenge itself becoming a source of entertainment. Sometimes these fan-made explanations become so popular and widespread that they slip into accepted canon, and the original creators of the characters accept them. For example, in the film *Return of the Jedi*, the character Boba Fett suffers a horrible death. However, the character was popular, so some fans held that he had somehow escaped "off-screen", and later books, graphic novels, and even an official action figure accepted this conjecture and depicted Boba Fett as having escaped the ordeal. In the commentary for the Special Edition Release of the film, George Lucas stated that had he known of the character's popularity, he would have made the death scene more impressive. Lucas left it ambiguous if the new interpretation was correct. However, in *Jedi Knight: Jedi Academy*, which takes place after Episode 6, the player must fight Boba Fett in Ord Mantel. (One reason Fett might have survived is he was drawn into the Sarlaac with all his weapons, unlike Jabba's other victims. It is not inconceivable he was able to use them to escape, killing the creature instead.)

It is commonplace for fictional characters appearing over a long period of time to remain the same age, or to age out of sync with real time. This concept, called a floating timeline, may be interpreted as an ongoing implicit retcon of their birthdate. When historical events are involved in their biography, overt retcons may be used to accommodate this; a character who served in the army during World War II might have his service record retconned to place him in the Korean War, the Vietnam War, the Gulf War, etc. A famous example of this type of retcon is the television show *The Simpsons* and Marvel Comics' characters *Nick Fury* and *The Punisher*. The *James Bond* movie series is another well-known example of this technique.

In live-action television series, real-world developments may prompt alteration-type retcons. For example, in *Star Trek: The Original Series*, limits in budget and technology resulted in the appearance of Klingons as swarthy-skinned people with vaguely central Asian features. When the franchise was later revived in films and new series enjoying larger budgets and improved makeup techniques, the appearance of Klingons was changed drastically. Skin looked more natural and spinal bones were brought up into the foreheads for a decidedly more alien appearance. The new look was explained by the producers to be how Klingons had always appeared, but that they could not be portrayed accurately before. The difference was marked upon in dialog between characters in "Trials and Tribble-ations", an episode of *Star Trek: Deep Space Nine* when modern-era characters travel back in time to the days of Kirk and Spock (and via then-brand new CGI techniques, appear within the TOS episode "The Trouble With Tribbles", frequently coming into contact with the *Enterprise* crew). When (modern) Klingon Worf and his crewmates see an original-series Klingon, his crewmates ask about the stunning difference in appearance; Worf says tersely that the matter is something Klingons "do not discuss with outsiders". This retcon itself was later retconned, in *Star Trek: Enterprise*, via a storyline in which it is revealed that the original, quasi-human appearance of the Klingons is due to a genetic mutation caused by an engineered virus – ironically, "genetic engineering" (Chief O'Brien) and "viral mutation" (Dr. Bashir) had been the guesses Worf refused to confirm or deny.

Subtraction

Unpopular or embarrassing stories are sometimes later ignored by publishers, never referred to again, and effectively erased from a series' continuity. The publishers may publish stories that contradict the previous story or explicitly establish that it "never happened" – for example by claiming that events in a previous installation were "just a dream", like one season of *Dallas*, which became Pam's dream so that her husband Bobby could return from the dead. An unpopular retcon may even be re-retconned away, as happened with John Byrne's *Spider-Man: Chapter One*.

An example of subtraction can be found in Disney's *The Lion King* series. After the success of the first movie, Disney released a group of books titled *The Lion King: Six New Adventures* in which Simba is said to have a son named Kopa. It is also mentioned in the storybook version of the film that he has a son. However, in the film sequel *The Lion King II: Simba's Pride*, Simba only has a daughter named Kiara. Kopa is non-existent and no mention is made of him. Kiara also has a different coloring and more feminine features than the cub shown at the end of the first movie.

This concept was parodied in *She-Hulk* (vol. 2) #3, in which the title character is threatened with erasure from continuity via a device that does such a thing. Eventually, she is pardoned, but not before two characters never seen before or since, Knight Man and Dr. Rocket, are erased from continuity, leaving those who didn't see it to respond, "Who?"

Reboot

Some retcons are used to erase all previous continuity and restart a fictional universe from the beginning. This type of retconning and its' effects are commonly referred to as a reboot. Reboots differ from retcons in that reboots have a built-in explanation – such as using time travel or reality warping to rewrite history – for the changes, while retcons are changes caused outside of the story by writers and/or editors. Some reboots – referred to as "soft" reboots – incorporate stories from the previous continuity into the new one rather than erasing them completely, while subjecting them to revision in order to fit them in.

The retcons in DC Comics' *Crisis on Infinite Earths* and *Flashpoint* storylines – both using the time travel method – were used to write an entirely new history for the DC Universe and its' characters. *Zero Hour* and *Infinite Crisis* used the reality warping method instead, resulting in minor changes to the DC Universe's history instead of a complete reboot.

The Spider-Man storyline *One More Day* – also using the reality warping method – culminated in the magical revision of history, eliminating the marriage between Peter Parker and Mary Jane Watson and subsequent developments in the character's history over the previous thirty years.

The Top Cow storyline *Artifacts* – also using the reality warping method – ended with the destruction and recreation of its fictional universe, with several continuity changes occurring as a result.

The "clean slate" reinterpretation of characters – as in movie and television adaptations of books, the reintroduction of superheroes in the Silver Age of Comics, or restarting a film series/television series/video game franchise from the beginning – also count as reboots, (and in rare cases include a built-in explanation for the changes, such as the time travel method used in the 2009 *Star Trek* film and the 2011 *Mortal Kombat* video game), except that the previous versions are not eliminated in the process. These are merely alternative reinterpretations, such as the character re-interpretations of the DC animated universe and the *Ultimate Marvel* line of comics.

Related

Retroactive continuity is similar to, but not the same as, plot inconsistencies introduced accidentally or through lack of concern for continuity; retconning is done deliberately. For example, the ongoing continuity contradictions on episodic TV series such as *The Simpsons* (in which the timeline of the family's history must be continually shifted forward to explain them not getting any older. For instance, when the series started, Bart would have to have been born in about 1980, but that would make him 30 years old as of 2010) reflects intentionally lost continuity, not genuine retcons. However, in series with generally tight continuity, retcons are sometimes created after the fact to explain continuity errors.

Retconning is also generally distinct from replacing the actor who plays a part in an ongoing series, which is more commonly an example of loose continuity rather than retroactively changing past continuity. The different appearance of the character is either ignored (as was done with the character of Darrin Stephens on the television show *Bewitched*, Vivian on *The Fresh Prince of Bel-Air*, or Claire on *My Wife and Kids*), or explained within the series, such as with "regeneration" in *Doctor Who*, or the Oracle in *The Matrix: Revolutions*.

It also differs from direct revision. For example, when George Lucas re-edited the original *Star Wars* trilogy, he made changes directly to the source material, rather than introducing new source material that contradicted the contents of previous material.

Literature involving retconning

When Sir Arthur Conan Doyle killed off his most-beloved character Sherlock Holmes by plunging him to his death over the Reichenbach Falls with his arch nemesis Professor Moriarty, the public's demand for Holmes was so great that Doyle was compelled to bring him back to life in a subsequent story, where he details that Holmes had merely faked his death.

In Stephen King's novel *Misery* the protagonist, Paul Sheldon, is forced to write a sequel to his book *Misery's Child*, in which the main character, Misery Chastain, dies. He at first attempts to retcon the events in that book, but his captor, Annie Wilkes, regards this as cheating and makes him create a sequel that doesn't actively deny what the reader already knows. The second attempt to bring Misery Chastain back to life (which Annie Wilkes likes) is almost an example of a comic book death.

Though the term "retcon" did not yet exist when George Orwell wrote *Nineteen Eighty-Four*, the totalitarian regime depicted in that book is involved in a constant, large-scale retconning of past records. For example, when it is suddenly announced that "Oceania was not after all in war with Eurasia. Oceania was at war with Eastasia and Eurasia was an ally" (Part Two, Ch. 9), there is an immediate intensive effort to change "all reports and records, newspapers, books, pamphlets, films, sound-tracks and photographs" and make them all record a war with Eastasia rather than one with Eurasia. "Often it was enough to merely substitute one name for another, but any detailed report of events demanded care and imagination. Even the geographical knowledge needed in transferring the war from one part of the world to another was considerable." See historical revisionism (negationism).

References in popular culture

The revived series of British science fiction television program *Doctor Who*, and its television spin-offs, heavily and playfully uses retroactive continuity plot devices. For example, in *Doctor Who* spin-off *Torchwood*, created by Russell T Davies, a drug used to erase the memory of characters is called "retcon"; the use of the drug is often referred to by characters as "retconning". The nod to retroactive continuity is a joke meant to be shared between the writers and the viewers as a way of pointing out that anything done throughout the course of the series can easily be undone with a simple plot device; it also points to parent show *Doctor Who*'s frequent use of the device in its several-decade run. For example, Davies introduced the Time War in the backstory to *Doctor Who*'s 2005 revival to account for discrepancies between the classic series and the revamp. When Steven Moffat took over in 2010, he introduced cracks in the universe which erase events and individuals from history; using this device, he 'undid' the events of "Journey's End" and "The Next Doctor" so that in the series' narrative, the people of Earth would be (once again) unaware of alien life. Moffat's fifth series finale provided a similar device when the Doctor "rebooted" the universe. In answer to a fan's question, Moffat tweeted: "The whole universe came exactly as it was. Except for any continuity errors I need to explain away."[9] And in the sixth series, Moffat introduces new aliens the Silence, who erase your memory of them the moment you look away. Creative use of the device is mined for new kinds of television suspense. In the episode "Day of the Moon", characters were shown to have had dozens of (unseen) encounters with the creatures in the space of a few seconds in viewer's time. Commenting on this device, writer MaryAnn Johanson writes, "That could be happening throughout this story... indeed, through the entire history of Doctor Who. Moffat has just created a pretty much unassailable narratively sound reason for inserting retcons anywhere throughout the half-century history of the show."[10]

The Daleks are another example of retconning in the *Doctor Who* series. Several times throughout the history of the show the Daleks have become extinct only to be 'reborn' at a later time due to their huge fan base and iconic placement as the Doctor's greatest foes.

In *Friendly Neighborhood Spider-Man*, the Hobgoblin of the year 2211 carries a weapon known as a 'Retcon Bomb'; upon impact, it erases its target and all memories of the target from existence, though not erasing the consequences of their existence (which is how the cracks in the skin of the universe work as well). This weapon has not been used since, because its inventor had fallen victim to one.

In *Champions Online*, a *MMORPG* for the PC, a character may use a "retcon" item to reset the statistics and powers of their superhero character.

One will also find frequent use of the term retcon in *Mystery Science Theater 3000* and the three shows spawned by it: *The Film Crew*, *Cinematic Titanic* and *Rifftrax*, all shows which mock films (usually incredibly bad ones).

The term retcon is used several times in the 2010 novel, *How to Live Safely in a Science Fictional Universe*, by American writer Charles Yu.

See also

- Back-story
- Chuck Cunningham Syndrome
- Cousin Oliver Syndrome
- Historical revisionism
- Jumping the shark
- List of retcons
- Plot hole
- Prequel
- Retcon (Torchwood)
- Retronym

Footnotes

[1] David, Peter. "Political Corrections, Part 2" (http://www.peterdavid.net/index.php/2010/09/20/political-corrections-part-2/) PeterDavid.net; September 20, 2010; Reprinted from *Comics Buyer's Guide* #1048, December 17, 1993

[2] Leith, Sam (2008-07-04). "Worshipping Doctor Who from behind the sofa" (http://www.telegraph.co.uk/comment/personal-view/3638405/One-of-these-comic-heroes-really-is-dead.html). *Daily Telegraph*. . Retrieved 2008-07-05.

[3] *The theology of Wolfhart Pannenberg* (http://books.google.com/books?ei=uQfSS8rpBcOqlAeT5LjtDA&ct=result&id=Rb0AAAAAMAAJ&dq="The+theology+of+Wolfhart+Pannenberg"&q=retroactive+continuity#search_anchor) excerpt at Google Books; Accessed September 20, 2010

[4] Thomas, Roy (w), Kubert, Joe (p), Hoberg, Rick (i). "Vengeance from Valhalla" *All-Star Squadron* 18 (February 1983), DC Comics

[5] "Original meaning of RETroactive CONtinuity" (http://groups.google.com/group/rec.arts.comics/msg/45793eff7c060796?dmode=source) at rec.arts.comics; Google Groups; Accessed September 20, 2010

[6] Moore, Alan (w), Bissette, Stephen R. (p), Totleben, John (i). "Anatomy Lesson" *Saga of the Swamp Thing* v2, 21 (February 1984), DC Comics

[7] Doyle, Arthur Conan (1893). "The Adventure of the Final Problem". *The Memoirs of Sherlock Holmes*.

[8] McAleer, John, *Rex Stout: A Biography*, pp. 403 and 566; see also *Over My Dead Body*

[9] "Steven Moffat: ...The whole universe..." (http://twitter.com/#!/steven_moffat/status/80157148893556736). *Steven Moffat on Twitter*. Twitter.com. 13 June 2011. . Retrieved 16 June 2011.

[10] "'Doctor Who' blogging: "Day of the Moon" | MaryAnn Johanson's" (http://www.flickfilosopher.com/blog/2011/05/doctor_who_blogging_day_of_the.html). Flickfilosopher.com. 2009-04-20. . Retrieved 2011-11-06.

External links

- Retconning: Just Another Day Like All The Others at websnark.com (http://www.websnark.com/archives/2008/01/retconning_just_1.html) An essay on the advantages and disadvantages of the various types of retcons.

AN/FSQ-31V

The **AN/FSQ-31V** was a computer made by IBM (International Business Machines) in 1960 and 1961 for the United States Air Force Strategic Air Command (SAC). The IBM Model name for the machine was the 4020[1] . Three **Q-31** units were built. They were used as the Data Processing Center (DPC) portion of the SAC Automated Command and Control System[2] .

Locations

Two (DPC 1 and DPC 2) were installed at SAC headquarters (Bldg 500) at Offutt Air Force Base outside Omaha, Nebraska. One (DPC 3) was installed in the Headquarters 15th Air Force Combat Operations Center at March Air Force Base, near Riverside, California. Another machine, designated the AN/FSQ-32, was installed at System Development Corporation (SDC) headquarters, Santa Monica, California and was used as a development machine for the compiler and operational software for the **Q-31**s.

Architecture

The system was divided into functional sections:

1. Central Processing Unit
2. Memory
3. High-Speed Input/Output
4. Low-Speed Input/Output
5. Operations Console

Central Processing Unit

Memory was addressed by words, which were 48 bits long. Each half word (24 bits) had a parity bit, for a total storage size of 50 bits. Depending on addressing mode, each word could be viewed as either 1 48-bit word, 2-24 bit half-words, 6 8-bit characters (EBCDIC encoded) or 8 6-bit bytes. A 6-bit byte, as opposed to the 8-bit byte in common use today, was common in IBM and other scientific computers of the time. The address space provided a maximum of 256K words.

The ISA was rather complicated for its time. The instructions were a fixed length of one word providing 24 bits for the operation and 24 bits for the address. The address consisted of 18 bits (3 bytes) for the memory address, with other bits used for the specification of index registers and indirect addressing.

The operation field provided the operation code and a variety of modifiers. Some modifiers allowed instructions to operate only on specific bytes of a word or on specific bits of a byte without separate masking operations. Other modifiers allowed the single 48-bit ALU to operate on a pair of 24-bit operands to facilitate vector operations.

Other parts of the CPU were some sense switches, which could be used to control various software functions, the run/halt switch, and a switch, amplifier, and speaker assembly, which could be used to provide audio feedback or even play music, by connecting one of four bits in the main accumulator which could then be toggled under software control at an appropriate rate to produce whatever tones one wanted. Someone in the software development division had produced a card deck on which was stored an executive that allowed selecting from a whole list of songs, including Christmas Carols, by setting the sense switches to a particular code, which would be printed out on the I/O Typewriter if a certain sense switch was on when the program was started.

Memory

The **Q-31**s were equipped with four 16 kiloword memory banks. The memory bank was oil and water cooled. Also considered as part of the memory subsystem in that they were addressed via fixed reserved memory addresses, were four 48 position switch banks, in which a short program could be inserted, and a plugboard, similar to the one used in IBM unit record equipment, that had the capacity of 32 words, so longer bootstrap or diagnostic programs could be installed in plug panels which could then be inserted into the receptacle and used. This served as a primitive ROM.

High-Speed Input/Output

The High-Speed I/O section provided interfaces to the Drum Memory system, which consisted of a control system, and two vertical drum memory devices. Each drum read and wrote 50 bits at a time in parallel so transferring data could be done quickly. The drums were organized as 17 fields with 8192 words per field for a total capacity of 139264 words. The motors that rotated the drums required 208 VAC at 45 Hz so a motor generator unit was required to change the frequency from 60 Hz. This added to the noise level in the computer room.

Low-Speed Input/Output

The Low-Speed I/O section interfaced to several different devices:

- SACCS EDTCC, which then interfaced to the rest of the SACCS system.
- Tape Controllers 1 and 2, connected to 16 IBM 729-V Tape Drives
- Disk File Controller, which was a modified Tape Controller, connected to the
  - Bryant PH 2000 Disk File, which had 24 disks that were 39 inches in diameter, 125 read/write heads that were hydraulically actuated, and had a total capacity of 26 MB
- IBM 1401, which controlled data transfers from unit-record equipment:
  - IBM 1402 Card Reader/Punch
  - IBM 1403 Line Printer
  - 2 IBM 729-V Tape Drives

- 2 IBM Selectric Typewriters, (I/O Typewriters) one of which was used for operational messages and the other for diagnostic messages and maintenance activities.

References

[1] "The IBM 4020 Military Computer" (http://bitsavers.org/pdf/ibm/4020/4020_Military_Computer_General_Info_Oct59.pdf). International Business Machines. 1959. . Retrieved September 28, 2009.

[2] Wohlman, John (1968). "Computer-Generated Map Data" (http://www.airpower.maxwell.af.mil/airchronicles/aureview/1968/jan-feb/wohlman.html). Air University Review. . Retrieved June 20, 2006.

External links

- Bitsavers.org IBM 4020 documentation (http://bitsavers.org/pdf/ibm/4020/)

Curtis_LeMay

Curtis LeMay	
[[File: |alt=]]	
Birth name	Curtis Emerson LeMay
Nickname	"Old Iron Pants", "Bombs Away" LeMay
Born	November 15, 1906 Columbus, Ohio, United States
Died	October 1, 1990 (aged 83) March Air Force Base, California, United States
Allegiance	United States
Service/branch	United States Air Force United States Army Air Forces United States Army Air Corps United States Army Ohio National Guard
Years of service	1928–65
Rank	General
Commands held	Twentieth Air Force Strategic Air Command USAF Chief of Staff
Battles/wars	World War II • Pacific Theatre
Awards	Distinguished Service Cross Army Distinguished Service Medal (3) Silver Star Distinguished Flying Cross (3) Air Medal (4) Distinguished Flying Cross (United Kingdom) Légion d'honneur (France) Grand Cordon, Order of the Rising Sun (Japan)
Other work	Politician, candidate for U.S. vice president

Curtis Emerson LeMay (November 15, 1906 – October 1, 1990) was a general in the United States Air Force and the vice presidential running mate of American Independent Party presidential candidate George Wallace in 1968.

He is credited with designing and implementing an effective, but also controversial, systematic strategic bombing campaign in the Pacific theater of World War II. During the war, he was known for planning and executing a massive bombing campaign against cities in Japan. After the war, he headed the Berlin airlift, then reorganized the Strategic Air Command (SAC) into an effective instrument of nuclear war.

LeMay became known for his massive incendiary attacks against Japanese cities during the war using hundreds of planes flying at low altitudes.

Early life and career

Curtis Emerson LeMay was born in Columbus, Ohio, on November 15, 1906. His father, Erving LeMay, was at times an ironworker and general handyman, but he never held a job longer than a few months. His mother, Arizona Carpenter LeMay, did her best to hold her family together. With very limited income, his family moved around the country as his father looked for work, going as far as Montana and California. Eventually they returned to his native city of Columbus. LeMay attended Columbus public schools, graduating from Columbus South High School, and studied civil engineering at Ohio State University. Working his way through college, he graduated with a bachelor's degree in civil engineering. While at Ohio State he was a member of the National Society of Pershing Rifles and the Professional Engineering Fraternity Theta Tau. He was commissioned a second lieutenant in the Air Corps Reserve in October 1929. He received a regular commission in the United States Army Air Corps in January 1930. While finishing at Ohio State, he took flight training at Norton Field in Columbus, in 1931–32.[1] On June 9, 1934, he married Helen E. Maitland (died 1992), with whom he had one child, Patricia Jane LeMay Lodge, known as Janie.

LeMay became a pursuit pilot and, while stationed in Hawaii, became one of the first members of the Air Corps to receive specialized training in aerial navigation. In August 1937, as navigator under pilot and commander Caleb V. Haynes on a Boeing B-17 Flying Fortress, he helped locate the battleship *Utah* despite being given the wrong coordinates by Navy personnel, in exercises held in misty conditions off California, after which the group of B-17s bombed it with water bombs. For Haynes again, in May 1938 he navigated three B-17s over 610 miles (**unknown operator: u'strong'** km) of the Atlantic Ocean to intercept the Italian liner *Rex* to illustrate the ability of land-based airpower to defend the American coasts. In 1940 he was navigator for Haynes on the prototype Boeing XB-15 heavy bomber, flying a survey from Panama over the Galapagos islands.[2] War brought rapid promotion and increased responsibility.

When his crews were not flying missions, they were being subjected to his relentless training, as he believed that training was the key to saving their lives. LeMay was widely and fondly known among his troops as "Old Iron Pants" throughout his career.[3]

World War II

When the U.S. entered World War II in December 1941, LeMay was a major in the United States Army Air Forces (he had been a 1st lieutenant as recently as 1940), and the commander of a newly created B-17 Flying Fortress unit, the 305th Bomb Group. He took this unit to England in October 1942 as part of the Eighth Air Force, and led it in combat until May 1943, notably helping to develop the combat box formation.[4] [5] He led the 4th Bombardment Wing and was its first commander when it was reorganized into the 3rd Air Division in September 1943. He personally led several dangerous missions, including the Regensburg section of the Schweinfurt-Regensburg mission of August 17, 1943. In that mission he led 146 B-17s to Regensburg, Germany, beyond the range of escorting fighters, and, after bombing, continued on to bases in North Africa, losing 24 bombers in the process.[4] [5] The heavy losses in veteran crews on this and subsequent deep penetration missions in the autumn of 1943 led the Eighth Air Force to limit missions to targets within escort range. Finally, with the deployment in the European theater of the P-51 Mustang in January 1944, the 8th Air Force gained an escort fighter with range to match the bombers.

Robert McNamara described LeMay's character, in a discussion of a report into high abort rates in bomber missions during World War II:

> One of the commanders was Curtis LeMay—Colonel in command of a B-24 group. He was the finest combat commander of any service I came across in war. But he was extraordinarily belligerent, many thought brutal. He got the report. He issued an order. He said, 'I will be in the lead plane on every mission. Any plane that takes off will go over the target, or the crew will be court-martialed.' The abort rate dropped overnight. Now that's the kind of commander he was.[6]

In August 1944, LeMay transferred to the China-Burma-India theater and directed first the XX Bomber Command in China and then the XXI Bomber Command in the Pacific. LeMay was later placed in charge of all strategic air operations against the Japanese home islands.[4] [5]

LeMay soon concluded that the techniques and tactics developed for use in Europe against the Luftwaffe were unsuitable against Japan. His bombers flying from China were dropping their bombs near their targets only 5% of the time. Operational losses of aircraft and crews were unacceptably high owing to Japanese daylight air defenses and continuing mechanical problems with the B-29. In January 1945 LeMay was transferred from China to relieve Brig. Gen. Haywood S. Hansell as commander of the XXI Bomber Command in the Marianas.[4] [5]

He became convinced that high-altitude precision bombing would be ineffective, given the usually cloudy weather over Japan. Furthermore, bombs dropped from the B-29s at high altitude (20,000+ feet) were often blown off of their trajectories by a consistently powerful jet stream over the Japanese home islands, which dramatically reduced the effectiveness of the high-altitude raids. Because Japanese air defenses made daytime bombing below jet stream-affected altitudes too perilous, LeMay finally switched to low-altitude nighttime incendiary attacks on Japanese targets, a tactic senior commanders had been advocating for some time.[4] [5] Japanese cities were largely constructed of combustible materials such as wood and paper. Precision high-altitude daylight bombing was ordered to proceed only when weather permitted or when specific critical targets were not vulnerable to area bombing. General Lemay was informed by a senior staff member, Colonel William P. Fisher, that bomber pilots were turning back from these low altitude bombing runs due to heavy anti-air craft fire from Japanese defense forces. Col. Fisher suggested to Lemay that crews who achieved successful strike rates should be rewarded by being released from their deployment. Lemay implemented this unorthodox plan and the strike rate went up to eighty percent.[7]

LeMay commanded subsequent B-29 Superfortress combat operations against Japan, including massive incendiary attacks on 64 Japanese cities. This included the fire-bombing of Tokyo on March 9–10, 1945, the most destructive bombing raid of the war.[8] For this first attack, LeMay ordered the defensive guns removed from 325 B-29s, loaded each plane with Model E-46 incendiary clusters, magnesium bombs, white phosphorus bombs, and napalm, and ordered the bombers to fly in streams at 5,000 to 9,000 feet over Tokyo.[4] [5]

The first pathfinder planes arrived over Tokyo just after midnight on March 10. Following British bombing practice, they marked the target area with a flaming "X." In a three-hour period, the main bombing force dropped 1,665 tons of incendiary bombs, killing some 100,000 civilians, destroying 250,000 buildings and incinerating 16 square miles (**unknown operator: u'strong'** km^2) of the city. Aircrews at the tail end of the bomber stream reported that the stench of burned human flesh permeated the aircraft over the target.[9]

A "LeMay Bombing Leaflet" from the war, which warned Japanese civilians that "Unfortunately, bombs have no eyes. So, in accordance with America's humanitarian policies, the American Air Force, which does not wish to injure innocent people, now gives you warning to evacuate the cities named and save your lives."

The New York Times reported at the time, "Maj. Gen. Curtis E. LeMay, commander of the B-29s of the entire Marianas area, declared that if the war is shortened by a single day, the attack will have served its purpose."[4] [5] Precise figures are not available, but the fire-bombing campaign against Japan, directed by LeMay between March 1945 and the Japanese surrender in August 1945, may have killed more than 500,000 Japanese civilians and left 5 million homeless.[10] Official estimates from the United States Strategic Bombing Survey put the figures at 220,000 people killed.[8] Some 40% of the built-up areas of 66 cities were destroyed, including much of Japan's war industry.[8]

The remaining Allied prisoners of war in Japan who had survived imprisonment to that time were frequently subjected to additional reprisals and torture after an air raid. The massive bombing also hit a number of prisons and directly killed a number of allied war prisoners. LeMay was quite aware of the Japanese opinion of him: he once remarked that had the U.S. lost the war, he fully expected to be tried for war crimes, especially in view of Japanese executions of uniformed American flight crews during the 1942 Doolittle raid. He argued that it was his duty to carry out the attacks in order to end the war as quickly as possible, sparing further loss of life.

Presidents Roosevelt and Truman justified these tactics by referring to an estimate of one million Allied casualties if Japan had to be invaded. The Japanese had intentionally decentralized 90 percent of their war-related production into small subcontractor workshops in civilian districts, making remaining Japanese war industry largely immune to conventional precision bombing with high explosives.[11]

As the fire-bombing campaign took effect, Japanese war planners were forced to expend significant resources to relocate vital war industries to remote caves and mountain bunkers, reducing production of war material. As an officer who served under LeMay, Lieutenant Colonel Robert McNamara was in charge of evaluating the effectiveness of American bombing missions. Later McNamara, as secretary of defense under Kennedy and Johnson, would often clash with LeMay.

LeMay also oversaw Operation Starvation, an aerial mining operation against Japanese waterways and ports that disrupted Japanese shipping and food distribution. Although his superiors were unsupportive of this naval objective, LeMay gave it a high priority by assigning the entire 313th Bombardment Wing (four groups, about 160 planes) to the task. Aerial mining supplemented a tight Allied submarine blockade of the home islands, drastically reducing Japan's ability to supply its overseas forces to the point that postwar analysis concluded that it could have defeated Japan on its own had it begun earlier.[4] [5]

Japan–Washington flight

LeMay piloted one of three specially modified B-29s flying from Japan to the U.S. in September 1945, in the process breaking several aviation records at that date, including the greatest USAAF takeoff weight, the longest USAAF non-stop flight, and the first ever non-stop Japan–Chicago flight. One of the pilots was of higher rank: Lieutenant General Barney Giles. The other two aircraft used up more fuel than LeMay's in fighting headwinds, and they could not fly to Washington, DC, the original goal.[12] They decided to land in Chicago to refuel. LeMay's aircraft had sufficient fuel to reach Washington, but he was directed by the War Department to join the others by refueling at Chicago. The order was ostensibly given because of borderline weather conditions in Washington, but according to First Lieutenant Ivan J. Potts who was aboard, the order came because LeMay had one fewer general's stars and should not be seen to outperform his superior.[13]

Cold War

Berlin Airlift

After World War II, LeMay was briefly transferred to The Pentagon as deputy chief of Air Staff for Research & Development. In 1947, he returned to Europe as commander of USAF Europe, heading operations for the Berlin Airlift in 1948 in the face of a blockade by the Soviet Union and its satellite states that threatened to starve the civilian population of the Western occupation zones of Berlin. Under LeMay's direction, C-54 cargo planes that could each carry 10 tons of cargo began supplying the city on July 1. By the fall, the airlift was bringing in an average of 5,000 tons of supplies a day. The airlift continued for 11 months—213,000 flights that brought in 1.7 million tons of food and fuel to Berlin. Faced with the failure of their blockade, the Soviet Union relented and reopened land corridors to the West. Though LeMay is sometimes publicly credited with the success of the Berlin Airlift, it was, in fact, instigated by General Lucius D. Clay when General Clay called LeMay about the problem. LeMay initially started flying supplies into Berlin, but then decided that it was a job for a logistics expert and he found that person in Lt. General William H. Tunner,[14] who took over the operational end of the Berlin Airlift.

General Curtis E. LeMay

Strategic Air Command

In 1948, he returned to the US to head the Strategic Air Command (SAC) at Offutt Air Force Base, replacing Gen George Kenney. When LeMay took over command of SAC, it consisted of little more than a few understaffed B-29 bombardment groups left over from WWII. Less than half of the available aircraft were operational, and the crews were undertrained. Base and aircraft security standards were minimal. Upon inspecting a SAC hangar full of US nuclear strategic bombers, LeMay found a single Air Force sentry on duty, armed only with a ham sandwich.[15] After ordering a mock bombing exercise on Dayton, Ohio, LeMay was shocked to learn that most of the strategic bombers assigned to the mission missed their targets by one mile or more. "We didn't have one crew, not one crew, in the entire command who could do a professional job"[16] noted LeMay.

In 1949, LeMay was first to propose that a nuclear war be conducted by delivering the nuclear arsenal in a single overwhelming blow, going as far as "killing a nation".[17]

In 1952 the 1st Missile Division was activated, having operational control over strategic missiles.

Upon receiving his fourth star in 1951 at age 44, LeMay became the youngest four-star general in American history since Ulysses S. Grant and was the youngest four-star general in modern history as well as the longest serving in that rank.[18] In 1956 and 1957 LeMay implemented tests of 24-hour bomber and tanker alerts, keeping some bomber forces ready at all times. LeMay headed SAC until 1957, overseeing its transformation into a modern, efficient, all-jet force. LeMay's tenure was the longest over an American military command in close to 100 years.[19]

Despite popular claims that LeMay advanced the notion of preventive nuclear war, the historical record indicates LeMay actually advocated justified preemptive nuclear war. Several documents show LeMay advocating preemptive attack of the Soviet Union, had it become clear the Soviets were preparing to attack SAC or the US. In these documents, which were often the transcripts of speeches before groups such as the National War College or events such as the 1955 Joint Secretaries Conference at the Quantico Marine Corps Base, LeMay clearly advocated using SAC as a preemptive weapon, if and when such was necessary.[20] Little evidence suggests LeMay ever advocated unauthorized or unjustified nuclear attack of the Soviet Union. To the contrary, a December 1949 letter from LeMay to the Air Force Chief of Staff Hoyt Vandenberg indicates LeMay was concerned with having explicit authority from the nation's political leadership to launch a preemptive strike against the Soviet Union. This letter, in LeMay's files at the Library of Congress, indicates LeMay was not willing to operate outside the authority afforded him as a military officer and that LeMay also recognized the constitutional role political leadership played in the decision to initiate war.

The "Airpower Battle"

General LeMay was instrumental in SAC's acquisition of a large fleet of new strategic bombers, establishment of a vast aerial refueling system, the formation of many new units and bases, development of a strategic ballistic missile force, and establishment of a strict command and control system with an unprecedented readiness capability. All of this was protected by a greatly enhanced and modernized security force, the Strategic Air Command Elite Guard. LeMay insisted on rigorous training and very high standards of performance for all SAC personnel, be they officers, enlisted men, aircrews, mechanics, or administrative staff, and reportedly commented, "I have neither the time nor the inclination to differentiate between the incompetent and the merely unfortunate."

A famous legend often used by SAC flight crews to illustrate LeMay's command style concerned his famous ever-present cigar.[21] In the first known published account of the story, *Life Magazine* reporter Ernest Havemann related that LeMay once took the co-pilot's seat of a SAC bomber to observe the mission, complete with lit cigar.[22] When asked by the pilot to put the cigar out, LeMay demanded to know why. When the pilot explained that fumes inside the fuselage could ignite the plane, LeMay reportedly growled, "It wouldn't dare."[22] The incident in the article was later used as the basis for a fictional scene in the 1955 film *Strategic Air Command.* In his highly controversial and factually disputed[23] [24] memoir *War's End*, Major General Charles Sweeney related an alleged 1944 incident that may have been the basis for the "It wouldn't dare" comment.[25]

Despite his uncompromising attitude regarding performance of duty, LeMay was also known for his concern for the physical well-being and comfort of his men.[26] LeMay found ways to encourage morale, individual performance, and the reenlistment rate through a number of means: encouraging off-duty group recreational activities,[27] [28] instituting spot promotions based on performance, and authorizing special uniforms, training, equipment, and allowances for ground personnel[29] as well as flight crews.

On LeMay's departure, SAC was composed of 224,000 airmen, close to 2,000 heavy bombers, and nearly 800 tanker aircraft.[19]

LeMay was an active amateur radio operator and held a succession of call signs; K0GRL, K4FRA, and W6EZV. He held these calls respectively while stationed at Offutt AFB, Washington, D.C. and when he retired in California. K0GRL is still the call sign of the Strategic Air Command Memorial Amateur Radio Club.[30] He was famous for being on the air on amateur bands while flying on board SAC bombers. LeMay became aware that the new single sideband (SSB) technology offered a big advantage over amplitude modulation (AM) for SAC aircraft operating long distances from their bases. In conjunction with Art Collins (W0CXX) of Collins Radio, he established SSB as the radio standard for SAC bombers in 1957.[31]

LeMay was appointed Vice Chief of Staff of the United States Air Force in July 1957, serving until 1961, when he was made the fifth Chief of Staff of the United States Air Force on the retirement of Gen Thomas White. His belief in the efficacy of strategic air campaigns over tactical strikes and ground support operations became Air Force policy during his tenure as chief of staff.

As chief of staff, LeMay clashed repeatedly with Secretary of Defense Robert McNamara, Air Force Secretary Eugene Zuckert, and the chairman of the Joint Chiefs of Staff, Army General Maxwell Taylor. At the time, budget constraints and successive nuclear war fighting strategies had left the armed forces in a state of flux. Each of the armed forces had gradually jettisoned realistic appraisals of future conflicts in favor of developing its own separate nuclear and nonnuclear capabilities. At the height of this struggle, the U.S. Army had even reorganized its combat divisions to fight land wars on irradiated nuclear battlefields, developing short-range atomic cannon and mortars in order to win appropriations. The United States Navy in turn proposed delivering strategic nuclear weapons from supercarriers intended to sail into range of the Soviet air defense forces. Of all these various schemes, only LeMay's command structure of SAC survived complete reorganization in the changing reality of postwar conflicts.

Though LeMay lost significant appropriation battles for the Skybolt ALBM and the B-52 Stratofortress replacement, the XB-70 Valkyrie, he was largely successful at expanding Air Force budgets. He advocated the introduction of satellite technology and pushed for the development of the latest electronic warfare techniques. By contrast, the U.S. Army and Navy frequently suffered budgetary cutbacks and program cancellations by Congress and Secretary McNamara.

During the Cuban Missile Crisis in 1962, LeMay clashed again with U.S. President John F. Kennedy and Defense Secretary McNamara, arguing that he should be allowed to bomb nuclear missile sites in Cuba. He opposed the naval blockade and, after the end of the crisis, suggested that Cuba be invaded anyway, even after the Russians agreed to withdraw. LeMay called the peaceful resolution of the crisis "the greatest defeat in our history".[32] Unknown to the US, the Soviet field commanders in Cuba had been given authority to launch—the only time such authority was delegated by higher command.[33] They had twenty nuclear warheads for medium-range R-12 ballistic missiles capable of reaching US cities (including Washington) and nine tactical nuclear missiles. If Soviet officers had launched them, many millions of US citizens would have been killed. The ensuing SAC retaliatory thermonuclear strike would have killed roughly one hundred million Soviet citizens, and brought nuclear winter to much of the Northern Hemisphere. Kennedy refused LeMay's requests, however, and the naval blockade was successful.[33]

The memorandum from LeMay, Chief of Staff, USAF, to the Joint Chiefs of Staff, January 4, 1964, illustrates LeMay's reasons for keeping bomber force along ballistic missiles: "It is important to recognize, however, that ballistic missile forces represent both the U.S. and Soviet potential for strategic nuclear warfare at the highest, most indiscriminate level, and at a level least susceptible to control. The employment of these weapons in lower level conflict would be likely to escalate the situation, uncontrollably, to an intensity which could be vastly disproportionate to the original aggravation. The use of ICBMs and SLBMs is not, therefore, a rational or credible response to provocations which, although serious, are still less than an immediate threat to national survival. For this reason, among others, I consider that the national security will continue to require the flexibility, responsiveness, and discrimination of manned strategic weapon systems throughout the range of cold, limited, and general war."[34]

LeMay's dislike for tactical aircraft and training backfired in the low-intensity conflict of Vietnam, where existing Air Force fighter aircraft and standard attack profiles proved incapable of carrying out sustained tactical bombing campaigns in the face of hostile North Vietnamese antiaircraft defenses. LeMay said, "Flying fighters is fun. Flying bombers is important."[35] Aircraft losses on tactical attack missions soared, and Air Force commanders soon realized that their large, missile-armed jet fighters were exceedingly vulnerable not only to antiaircraft shells and missiles but also to cannon-armed, maneuverable Soviet fighters.

LeMay advocated a sustained strategic bombing campaign against North Vietnamese cities, harbors, ports, shipping, and other strategic targets. His advice was ignored. Instead, an incremental policy was implemented that focused on limited interdiction bombing of fluid enemy supply corridors in Vietnam, Laos, and Cambodia. This limited campaign failed to destroy significant quantities of enemy war supplies or diminish enemy ambitions. Bombing limitations were imposed by President Lyndon Johnson for geopolitical reasons, as he surmised that bombing Soviet and Chinese ships in port and killing Soviet advisers would bring the Soviets more directly into the war and destabilize the European Cold War.

Evidence of LeMay's thinking is that in his 1965 autobiography, co-written with MacKinlay Kantor, LeMay is quoted as saying his response to North Vietnam would be to demand that "they've got to draw in their horns and stop their aggression, or we're going to bomb them back into the Stone Age. And we would shove them back into the Stone Age with Air power or Naval power—not with ground forces."

Some military historians have argued that LeMay's theories were eventually proven correct. Near the war's end in December 1972, President Richard Nixon ordered Operation Linebacker II, a high-intensity Air Force, Navy, and Marine Corps aerial bombing campaign, which included hundreds of B-52 bombers that struck previously untouched North Vietnamese strategic targets, including heavy populated areas in Hanoi and Haiphong. Linebacker II, despite being one of the most intense bombing operations of the conflict, greatly weakened but did not cause the collapse of the communist regime, although it forced the communist government back into peace talks and shortly after the operation was called off. The Paris Peace Agreement was signed, and some of it clearly benefited the government of North Vietnam. During the operation, the U.S Air Force and Navy lost 17 strategic bombers.

Post-military career

Owing to his unrelenting opposition to the Johnson administration's Vietnam policy and what was widely perceived as his hostility to Secretary McNamara, LeMay was essentially forced into retirement in February 1965 and seemed headed for a political career. Moving to California, he was approached by conservatives to challenge moderate Republican Thomas Kuchel for his seat in the United States Senate in 1968, but he declined. For the presidential race that year, LeMay originally supported Richard Nixon; he turned down two requests by George Wallace to join his American Independent Party that year on the grounds that a third-party candidacy might hurt Nixon's chances at the polls (by coincidence, Wallace had served as a sergeant in a unit commanded by LeMay during WWII). LeMay gradually became convinced that Nixon planned to pursue a conciliatory policy with the Soviets and accept nuclear parity rather than retain America's first-strike supremacy. LeMay felt that Lyndon Johnson had lied to him on several occasions, and Hubert Humphrey, if elected, would do the same. Consequently LeMay, while being fully aware of Wallace's segregationist platform, decided to throw his support to Wallace and eventually became Wallace's running mate. The general was dismayed to find himself attacked in the press as a racial segregationist because he was running with Wallace; he had never considered himself a bigot. When Wallace announced his selection in October 1968, LeMay opined that he, unlike many Americans, clearly did not fear using nuclear weapons. His saber rattling did not help the Wallace campaign.

During the 1968 campaign, LeMay became widely associated with the "Stone Age" comment, especially because he had suggested use of nuclear weapons as a strategy to quickly resolve a deeply protracted conventional war which eventually claimed over 50,000 American and millions of Vietnamese lives. This reputation did nothing to diminish perceptions of extremism in the Wallace-LeMay ticket. General LeMay disclaimed the comment, saying in a later interview: "I never said we should bomb them back to the Stone Age. I said we had the capability to do it."

The Wallace-LeMay AIP ticket received 13.5 percent of the popular vote, higher than most third-party candidacies in the US, and carried 5 states for a total of 46 electoral votes, but this was not enough to deny Nixon his election as 37th President of the US. It was enough to stop Hubert Humphrey, and Nixon's ground breaking success as a Republican winning electoral votes from Democratic southern states was partly due to Democrats not voting for the Democratic ticket. Following the 1968 election, LeMay returned to private life, including pursuing several charitable projects. He declined further suggestions to run for political office.

He was honored by several countries, receiving the Air Medal with three oak leaf clusters, the Distinguished Flying Cross with two oak leaf clusters, the Distinguished Service Cross, Distinguished Service Medal with two oak leaf clusters, the French Legion of Honor and the Silver Star. On December 7, 1964 the Japanese government conferred on him the First Order of Merit with the Grand Cordon of the Order of the Rising Sun. He was elected to the Alfalfa Club in 1957 and served as a general officer for 21 years.

According to historian Warren Kozak, Wallace's defeat left LeMay's public reputation in tatters. LeMay was commonly assumed to share Wallace's widely unpopular racist views, even though LeMay had enthusiastically supported racial integration in the US military publicly and privately. He fought segregation in the Air Force before Executive Order 9981 systemically banned the practice.[36]

Death

General LeMay died on October 1, 1990, at March Air Force Base in Riverside County, California, and is buried in the United States Air Force Academy Cemetery[37] at Colorado Springs, Colorado. He was survived by his wife Helen, who died in 1992 and is buried next to the general.

Miscellany

LeMay and UFOs

The April 25, 1988 issue of *The New Yorker* carried an interview with retired Air Force Reserve Major General and former US Senator from Arizona, Barry Goldwater, who said he repeatedly asked his friend General LeMay if he (Goldwater) might have access to the secret "Blue Room" at Wright Patterson Air Force Base, alleged by numerous Goldwater constituents to contain UFO evidence. According to Goldwater, an angry LeMay gave him "holy hell" and said, "Not only can't you get into it but don't you ever mention it to me again."[38]

LeMay and sports car racing

General LeMay was also a sports car owner and enthusiast (he owned an Allard J2); as the "SAC era" began to wind down, LeMay loaned out facilities of SAC bases for use by the Sports Car Club of America,[39] as the era of early street races began to die out. He was awarded the Woolf Barnato Award, SCCA's highest award for contributions to the Club, in 1954.[39] In November 2006, it was announced that General LeMay would be one of the inductees into the SCCA Hall of Fame in 2007.[39]

Lemay and the Randall Firearms Company

In 1984, as a result of the friendship between Lemay and the company's chief designer, Randall began producing the first of what would be two different models of compact .45 ACP pistol with the Lemay 4-Star designation.[40]

Air Force Academy Exemplar

On March 13, 2010, LeMay was named the exemplar for the United States Air Force Academy class of 2013.[41]

Rank history

Curtis LeMay's first contact with military service occurred in September 1924 when he enrolled as a student in the ROTC program at Ohio State University. By his senior year, LeMay was listed on the ROTC rolls as a "cadet lieutenant colonel" but had not actually received an appointment in the regular United States military.

On June 14, 1928, the summer before the start of his senior year, LeMay accepted a commission as a second lieutenant in the Field Artillery Reserve of the United States Army. In September 1928, LeMay was approached by the Ohio National Guard and asked to accept a state commission, also as a second lieutenant, which LeMay accepted. This created a unique situation in LeMay's service record since in 1928 it was unusual for a person to hold a commission both in the National Guard and the Army Reserve.

On September 29, 1928, LeMay enlisted in the Army Air Corps as an aviation cadet under the service number 6650359. For the next 13 months, LeMay was not only on the enlisted rolls of the Regular Army but also still held a commission in the National Guard and Army Reserve. Thus, for this short period in LeMay's career, he was technically an officer and enlisted soldier at the same time, a practice no longer permitted in the U.S. military. The matter was resolved on October 2, 1929, when LeMay's Guard and Reserve commission were terminated. According to his service record, these commissions were revoked "by telephone" after an Army personnel officer, realizing that LeMay was holding officer and enlisted status simultaneously, called him to discuss the matter.

On October 12, 1929, LeMay finished his flight training and was commissioned a second lieutenant in the Air Corps Reserve. This was the third time he had been appointed a second lieutenant in just under two years. He held this reserve commission until June 1930, when he was appointed as a Regular Army officer in the Air Corps.

LeMay experienced slow advancement throughout the 1930s, as did most officers of the seniority-driven regular army. At the start of 1940 he was still a first lieutenant but, beginning in 1941, began to receive temporary advancement in grade in the expanding Army Air Forces. LeMay advanced from captain to brigadier general in less than four years and by 1944 was a major general. When World War II ended, he was appointed to the permanent rank of brigadier general in the Regular Army but held his temporary rank of major general in the Army until promotion to lieutenant general in the now separate United States Air Force in 1948. He then was promoted to full general in 1951 and held this rank until his retirement in 1965.

Dates of rank

- Army ROTC Cadet: September 1924
- Second Lieutenant, Field Artillery Reserve: June 14, 1928
- Second Lieutenant, Ohio National Guard: September 22, 1928
- Flight Cadet, Army Air Corps: September 28, 1928
- Officer Commissions Terminated: October 2, 1929
- Second Lieutenant, Air Corps Reserve: October 12, 1929
- Second Lieutenant, Army Air Corps: February 1, 1930
- First Lieutenant, Army Air Corps: March 12, 1935
- Captain, Army Air Corps: January 26, 1940
- Major, Army Air Corps: March 21, 1941
- Lieutenant Colonel, Army of the United States: January 23, 1942
- Colonel, Army of the United States: June 17, 1942
- Brigadier General, Army of the United States: September 28, 1943
  - Permanent in the Regular Army: June 22, 1946
- Major General, Army of the United States: March 3, 1944
- Lieutenant General, United States Air Force: January 26, 1948
- General, United States Air Force: October 29, 1951
- General, USAF (Retired): February 1, 1965

Further promotion

According to letters in Curtis LeMay's service record, while he was in command of SAC during the 1950s several petitions were made by Air Force service members to have LeMay promoted to the rank of General of the Air Force (five stars). The Air Force leadership, however, felt that such a promotion would lessen the prestige of this rank, which was seen as a wartime rank to be held only in times of extreme national emergency.

Per the Chief of the Air Force General Officers Branch, in a letter dated February 28, 1962:

> It is clear that a grateful nation, recognizing the tremendous contributions of the key military and naval leaders in World War II, created these supreme grades as an attempt to accord to these leaders the prestige, the clear-cut leadership, and the emolument of office befitting their service to their country in war. It is the conviction of the Department of the Air Force that this recognition was and is appropriate. Moreover, appointments to this grade during periods other than war would carry the unavoidable connotation of downgrading of those officers so honored in World War II.

Thus, no serious effort was ever made to promote LeMay to the rank of General of the Air Force, and the matter was eventually dropped after his retirement from active service in 1965.

Awards and decorations

LeMay received recognition for his work from thirteen countries, receiving twenty-two medals and decorations.

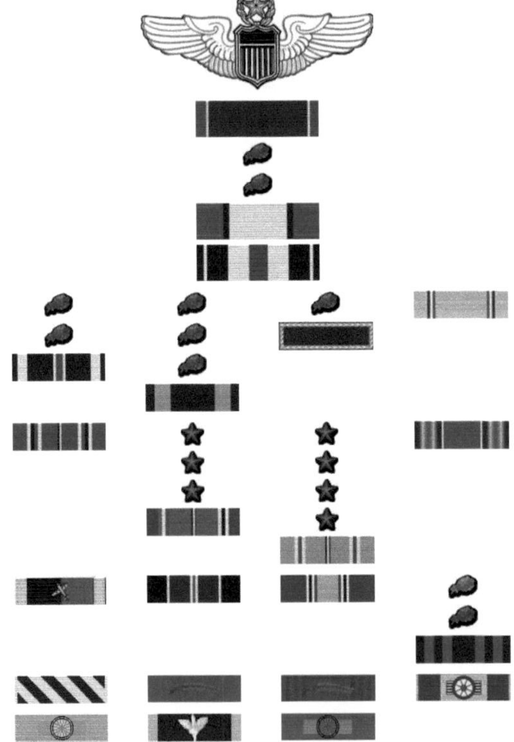

United States

- Command pilot
- Distinguished Service Cross
- Distinguished Service Medal plus 2 oak leaf clusters
- Silver Star
- Distinguished Flying Cross plus 2 oak leaf clusters
- Air Medal plus 3 oak leaf clusters
- Presidential Unit Citation plus oak leaf cluster
- American Defense Service Medal
- American Campaign Medal
- European-African-Middle Eastern Campaign Medal plus 3 bronze campaign stars
- Asiatic-Pacific Campaign Medal plus 4 bronze campaign stars
- World War II Victory Medal
- Army of Occupation Medal with Airlift Device
- Medal for Humane Action
- National Defense Service Medal
- Air Force Longevity Service Award, 6 oak leaf clusters

Other countries

- British Distinguished Flying Cross
- French Croix de Guerre with Palm
- Belgian Croix de Guerre, with Palm
- Japanese Order of the Rising Sun, Grand Cordon
- Brazilian Order of the Southern Cross
- Brazilian Order of Aeronautical Merit
- Moroccan Order of Ouissam Alaouite
- Swedish Commander Grand Cross of the Royal Order of the Sword
- Argentina – Order of Aeronautical Merit – Grades of Grand Official and Grand Cross
- Chile – Order of the Merit
- Chile – Medalla Militar de Primera Clase
- Ecuador – Order of Aeronautical Merit (Knight Commander)
- Uruguay – Aviador Militar Honoris Causa (Piloto Commandante)
- U.S.S.R – Order of the Patriotic War – 1st Degree

Works

Books

- LeMay, Curtis; Kantor, MacKinlay (1965), *Mission with LeMay: My Story*, Doubleday, B00005WGR2.
- LeMay, Curtis; Smith, Dale O (1968), *America is in Danger*, Funk & Wagnalls, B00005VCVX.
- LeMay, Curtis; Yenne, William 'Bill' (1988), *Superfortress: The Story of the B-29 and American Air Power*, McGraw-Hill, ISBN 0-07-037160-1.

LeMay, Curtis, Warren Kozak, The Life and Wars of General Curtis LeMay, Regnery Publishing Inc, 2009, ISBN 978-1-59698-569-8

Film and television appearances

- *The Last Bomb* (documentary, 1945)
- *In the Year of the Pig* (documentary, 1968)
- *The World at War* (documentary TV series, 1974)
- *Race for the Superbomb* (documentary, 1999)
- *JFK* (film, 1991; featured in archival footage)
- *Roots of the Cuban Missile Crisis* (documentary, 2001)
- *The Fog of War: Eleven Lessons from the Life of Robert S. McNamara* (documentary, 2003)
- *DC3:ans Sista Resa* (Swedish documentary, 2004)

In popular culture

- *Above and Beyond* – LeMay is portrayed by Jim Backus (film, 1952)
- *Strategic Air Command* – the character of General Ennis C. Hawkes, based on LeMay, is played by Frank Lovejoy (film, 1955)
- *Dr. Strangelove or: How I Learned to Stop Worrying and Love the Bomb* – the character of General Buck Turgidson, played by George C. Scott, is based in part on LeMay (film, 1964)[42]
- *The Missiles of October* – LeMay is played by Robert P. Lieb (TV, 1974)[43]
- *Enola Gay: The Men, the Mission, the Atomic Bomb* – LeMay is portrayed by Than Wyenn (TV, 1980)
- *Kennedy* – played by Barton Heyman (TV series, 1983)[43]

- *Race for the Bomb* – played by Lloyd Bochner (TV series, 1987)[43]
- *Hiroshima* played by Cedric Smith (TV, 1995)[43]
- *Thirteen Days* – LeMay is played by Kevin Conway (film, 2000)[43]
- *Roots of the Cuban Missile Crisis* – played by Kevin Conway (video, 2001)[43]
- *Black Wind by F. Paul Wilson (fiction), in which LeMay appears in connection with the Hiroshima bombing.*[44]

Public buildings

- The headquarters building of U.S. Strategic Command at Offutt AFB in Nebraska is named for the general.
  It was erected in the late 1950s and was the headquarters of the Strategic Air Command until its disbandment in 1992.
- LeMay Elementary School opened in 1968 in the Capehart housing area of Offutt AFB and is operated by the Bellevue Public Schools.[45]

Gen. Curtis E. LeMay Building,
U.S. Strategic Command Headquarters

References

Notes

[1] *Ohio History Central* (http://www.ohiohistorycentral.org/entry.php?rec=773), .

[2] Boniface, Patrick (Jan–Feb. 1999), "Boeing's Forgotten Monster: XB-15, a Giant in Search of a Cause", *Air Enthusiast*, pp. 64–7.

[3] Harper, CB (Red). "March 1944 and Berlin" (http://www.buffalogal.org/MarchBerliln.htm). *With The Mighty Eighth And The Fifteenth Air Forces In Action Over Europe In World War II.* .

[4] Coffey, *Iron Eagle*

[5] Tillman, *LeMay*

[6] Errol Morris, *The Fog of War: Eleven Lessons from the Life of Robert S. McNamara,* Documentary Film, 2003

[7] December 24, 1985 recorded interview with Major General William P. Fisher, USAF, Retired. Conducted by granddaughter Dorothy Danaher White

[8] United States Strategic Bombing Survey. Summary Report (Pacific War). Washington DC, July 1, 1946.

[9] Buckley, John (2001) [1998]. *Air Power in the Age of Total War.* London: Taylor & Francis. p. 193 (http://books.google.co.uk/books?id=0_-DJcwmQ20C&pg=PA193&dq="many+wore+their+oxygen+masks+to+escape+the+sickening+stench+of+burning+flesh"). ISBN 0-203-00722-0.

[10] Bradley, F. J. *No Strategic Targets Left.* "Contribution of Major Fire Raids Toward Ending WWII", Turner Publishing Company, limited edition. ISBN 1-56311-483-6. p. 38.

[11] John Toland, *The Rising Sun: The Decline and Fall of the Japanese Empire 1936–1945,* Random House, 1970, p. 671.

[12] *40th Bombardment Group (VH) history* (http://books.google.com/books?id=EB-0VUI11NsC&pg=PA46). Turner Publishing. 1989. pp. 45–47. ISBN 0-938021-28-1. .

[13] Potts, J. Ivan, Jr. "The Japan to Washington Flight: September 18–19, 1945" (http://www.40thbombgroup.org/DCFlight.pdf) (PDF). 40th Bomb Group. . Retrieved October 19, 2010.

[14] Cherny, Andrei, *The Candy Bombers: The Untold Story of the Berlin Airlift and America's Finest Hour*, Putnam Press, ISBN 978-0-399-15496-6 (2008)

[15] Watson, George M., *Secretaries and Chiefs of Staff of the United States Air Force*, Washington, D.C.: Air Force History and Museums Program, USAF (2001) p. 132: LeMay recorded the incident in a memo to staff the same day, stating "this afternoon I found a man guarding a hangar with a ham sandwich. There will be no more of that."

[16] Ford, Daniel (April 1, 1996). "History of Flight – B-36: Bomber at the Crossroads" (http://www.airspacemag.com/history-of-flight/B-36-Bomber-at-the-Crossroads.html?c=y&page=5). *Air & Space Magazine*. .

[17] DeGroot, Gerard J. (2004). *The bomb: a life* (http://books.google.com/books?id=6VQCsAZpPrgC&lpg=PA153&dq=the entire stockpile of atomic bombs in a single massive attack&pg=PA153#v=onepage&q&f=false) (1st, pbk. ed.). Cambridge, MA: Harvard University Press. p. 153. ISBN 0-674-01724-2. .

[18] Kozak, Warren. "LeMay: The Life And Wars Of General Curtis LeMay" (http://www.warrenkozak.com/basics.html). . Retrieved 2009-05-08.

[19] *AIR FORCE Magazine*. October 2008.

[20] Library of Congress, Manuscript Reading Room, Papers of Curtis E. LeMay, Box B-205, Remarks by General Curtis E. LeMay at Quantico, 15 July 1955.

[21] Havemann, Ernest, *Toughest Cop of The Western World* (http://books.google.com/books?id=HFMEAAAAMBAJ&pg=PA132&lpg=PA132&dq=LeMay+Toughest+Cop+of+the+Western+World&source=bl&ots=XiRLy7LlDr&sig=pQNJpzP0Gcjgk7gchsjwcTz6azU&hl=en&ei=zDyfTcyNBOrjiAKUnLHzAg&sa=X&oi=book_result&ct=result&resnum=1&ved=0CBgQ6AEwAA), Life Magazine, 14 June 1954, p. 136

[22] Havemann, p. 136

[23] Puttré, Michael, *Nagasaki Revisited* (http://www.mputtre.com/id45.html), retrieved 8 April 2011

[24] Coster-Mullen, John, *Atom Bombs: The Top Secret Inside Story of Little Boy and Fat Man,* publ. J. Coster-Mullen, End Notes (2004): Gen. Paul Tibbets, Major Dutch Van Kirk (Enola Gay's navigator), and other surviving members of the 509th Composite Group were reportedly outraged at many of the factual assertions by Sweeney in *War's End.*

[25] Sweeney, Charles (Maj. Gen., ret.), Antonucci, James A., and Antonucci, Marion K., *War's End: an Eyewitness Account of America's Last Atomic Mission*, New York: Avon Books, ISBN 0-380-97349-9 (1997), p. 75: Sweeney stated that a similar incident occurred in 1944 when a B-29 crew chief reminded General LeMay of his lit cigar while LeMay was undergoing B-29 familiarization with (then-Colonel) Paul Tibbets' 509th Composite Group.

[26] Watson, George M., *Secretaries and Chiefs of Staff of the United States Air Force*, Washington, DC: Air Force History and Museums Program, USAF (2001) p. 132.

[27] *Sport: Red for Ferrari*, Time Magazine, 20 April 1953.

[28] *Judo In SAC Air Force, Black Belt Magazine*, April 1962, pp. 37–38: These ranged from basketball courts and pool tables to judo tournaments and even assembling and tuning engines in SAC workshops for sports car races on SAC air bases.

[29] *Armed Forces: The Finish Flag, Time Magazine*, 2 August 1954: This included new innerspring mattresses, fans, pool tables, and TV sets for enlisted mens quarters.

[30] "Surfin': More Hamming at 1600 Pennsylvania Avenue" (http://www.arrl.org/news/features/2005/02/18/1/). National Association for Amateur Radio. .

[31] "Amateur Radio and the Rise of SSB" (http://www.arrl.org/qst/2003/01/McElroy.pdf) (PDF). National Association for Amateur Radio. .

[32] Feltus, Pamela. "Curtis E. LeMay" (http://www.centennialofflight.gov/essay/Air_Power/LeMay/AP36.htm). *Centennial of Flight.* US Centennial of Flight Commission. . Retrieved 10 April 2010.

[33] Rhodes, 1995, pp. 574–76

[34] National Archives and Records Administration, RG 200, Defense Programs and Operations, LeMay's Memo to President and JCS Views, Box 83. Secret.

[35] Robert Coram, "Boyd. *Back Bay Books/Little, Brown, and Company, 2002, p. 59.*

[36] Glazov, James 'Jamie' (2009-10-6), "LeMay: the life and wars of General Curtis LeMay" (http://frontpagemag.com/2009/10/06/lemay-the-life-and-wars-of-general-curtis-lemay-by-jamie-glazov/), *Front Page Magazine*, .

[37] "Curtis Emerson LeMay" (http://www.findagrave.com/cgi-bin/fg.cgi?page=gr&GRid=9654). Find-A-Grave. . Retrieved June 22, 2010.

[38] *The New Yorker* (April 25, 1988)

[39] "SCCA Announces 2007 Hall of Fame Class" (http://web.archive.org/web/20061205225745/http://www.scca.com/News/News.asp?Ref=729). Sports Car Club of America. November 22, 2006. Archived from the original (http://www.scca.com/News/News.asp?Ref=729) on 2006-12-05. .

[40] url=http://www.sightm1911.com/lib/history/randall_history.htm

[41] *Class exemplar* (http://69.199.231.171/wiki/index.php/Class_exemplar), .

[42] Carter, Dan T. (1995). *The Politics of Rage: George Wallace, the Origins of the New Conservatism, and the Transformation of American Politics*. New York: Simon & Schuster. p. 357. ISBN 0-8071-2597-0.

[43] "General Curtis LeMay (character)" (http://www.imdb.com/character/ch0177916/) on Internet Movie Database

[44] Wilson, F. Paul (2009). *Black Wind.* Tor Books. ISBN 0-7653-6292-9. (http://www.repairmanjack.com/forum/content.php?71-Black-Wind)

[45] "LeMay Elementary" (http://www.bellevuepublicschools.org/lemay/index.cfm?action=202&id=423&tab=8), *Bellevue public schools*, , retrieved 2011-09-08.

Bibliography

- Atkins, Albert *Air Marshall Sir Arthur Harris and General Curtis E. Lemay: A Comparative Analytical Biography*. AuthorHouse, 2001. ISBN 0-7596-5940-0.
- Craig, William *The Fall of Japan*. The Dial Press, 1967. Library of Congress Catalog Card No. 67-10704.
- Coffey, Thomas M. *Iron Eagle: The Turbulent Life of General Curtis LeMay*. Random House, 1986. ISBN 0-517-55188-8.
- Kozak, Warren *LeMay: The Life and Wars of Curtis LeMay*. Regnery, 2009.
- LeMay, Curtis E. "Mission with LeMay: My Story". Doubleday, 1965
- LeMay, Curtis E., Yenne, Bill *Superfortress: The Boeing B-29 and American Airpower in World War II*. Westholme Publishing 2006, originally published by Berkley, 1988
- McNamara, Robert S. *In Retrospect: The Tragedy and Lessons of Vietnam*. Vintage Press, 1995. ISBN 0-679-76749-5.
- Moscow, Warren "City's Heart Gone". *The New York Times*. March 11, 1945: 1, 13.
- Narvez, Alfonso A. "Gen. Curtis LeMay, an Architect of Strategic Air Power, Dies at 83". *The New York Times*. October 2, 1990.
- Allison, Graham. *Essence of Decision*: Explaining the Cuban Missile Crisis (1971 – updated 2nd edition, 1999). Longman. ISBN 0-321-01349-2.
- Rhodes, Richard *Dark Sun: The Making of the Hydrogen Bomb*. Simon & Schuster, 1995. ISBN 0-684-80400-X
- Tillman, Barrett. *LeMay*. Palgrave's Great Generals Series, 2007. ISBN 1-4039-7135-8
- USAF National Museum, "Gen. Curtis E. LeMay, Awards and Decorations" (http://www.nationalmuseum.af.mil/factsheets/factsheet.asp?id=1116)
- USAF Service Record of Curtis LeMay, Military Personnel Records Center, St. Louis, MO
- "Curtis E LeMay" (http://www.456fis.org/CURTIS_E._LAMAY.htm), *456 Fighter Interceptor Squadron*.

External links

- "LeMay" (http://www.afa.org/magazine/March1998/0398lemay.asp), *Air Force Magazine*, USA, March 1998.
- Meilinger, Colonel Phillip S, "LeMay" (http://www.airpower.maxwell.af.mil/airchronicles/cc/lemay.html), *American Airpower Biography: A Survey of the Field*, USAF.
- Rhodes, Richard, *General Curtis LeMay, Head of Strategic Air Command* (http://www.pbs.org/wgbh/amex/bomb/filmmore/reference/interview/rhodes06.html), USA: PBS.
- Rhodes, Richard, *LeMay's Vision of War* (http://www.pbs.org/wgbh/amex/bomb/filmmore/reference/interview/rhodes07.html), USA: PBS.
- *Annotated bibliography of Curtis LeMay* (http://alsos.wlu.edu/qsearch.aspx?browse=people/LeMay,+Curtis), Alsos Digital Library.
- "Gen. Curtis E. LeMay" (http://www.nationalmuseum.af.mil/factsheets/factsheet.asp?id=1115), *National Museum*, USAF.
- Nutter, Ralph H, *With the Possum and the Eagle* (http://books.google.com/books?id=v3D0_vLaPu8C&pg=PA7&lpg=PA7&dq="navigator's+wings&source=bl&ots=lt341XM21N&sig=YHRLEI9z95-_Qh_LqQ8_Tu5CJZ0&hl=en&sa=X&oi=book_result&resnum=7&ct=result#PPP1,M1http://books.google.com/books?id=v3D0_vLaPu8C&pg=PA7&lpg=PA7&dq="navigator's+wings&source=bl&ots=lt341XM21N&sig=YHRLEI9z95-_). A view of working with LeMay, by his lead navigator in Europe.
- "Curtis LeMay" (http://www.findagrave.com/cgi-bin/fg.cgi?page=gr&GRid=9654). Find a Grave. Retrieved 2008-02-17.

Article Sources and Contributors

Strategic_Automated_Command_and_Control_System *Source*: http://en.wikipedia.org/w/index.php?title=Strategic_Automated_Command_and_Control_System *Contributors*: A mimic justly wanes, Brenont, Centrx, Cgingold, ChardingLLNL, Chuckharding, D6, DocYako, Edward, L'amateur d'aéroplanes, Mais oui!, Ndunruh, Paul A. Romsky Jr., PigFlu Oink, Rich Farmbrough, Tdrss, Wa3frp, 9 anonymous edits

Intercontinental_ballistic_missile *Source*: http://en.wikipedia.org/w/index.php?title=Intercontinental_ballistic_missile *Contributors*: .:Ajvol:., 128.59.58.xxx, 1exec1, Abanamat, Abcx123, Abhishek191288, AceKingQueenJack, Aditya, Adzze, Ale jrb, Alex Cohn, AlexiusHoratius, Alfio, Ali 786, Alice Muller, Alinor, Allens, Alro, Altenmann, Amadoraa123, Amorrow, Anatidae, Anir1uph, Antares Geminorum, Anthony Appleyard, Ao333, Aogouguo, Apocalyptic Destroyer, Apyule, Arado, Arcyqwerty, Arka Voltchek, Asd36f, Ash sul, AssiPunjabi, Assie Travis, Astronautics, Atomicgurl00, Atomique, Audin, Auspiv, Avs5221, B21O303V3941W42371, BD2412, BRPXQZME, Balcer, Bamac30, Bambuway, Bcs09, Benandorsqueaks, Bentogoa, Bharatgopal, Bijal d g, BlaiseFEgan, Bobblehead, Bobdobbs1723, Briaboru, Brighterorange, Bryan Derksen, Burto88, By78, CKC01, Cal 1234, Can't sleep, clown will eat me, Catgut, Chagai, Chanakyathegreat, Chessofnerd, Chinfo, Chopduel, Chuggsymalone, Cmskog, Colonies Chris, Coma28, Conversion script, Cosmos416, Craigboy, CranialNerves, Crispmuncher, Cumulus, Cunikm, D, Dabomb87, Damuna, Dan100, DanMS, Daniel C. Boyer, Dbchip, Deathmare, Deavenger, Debastein, Dgw, Dherky, Digitalsuresh, Dijxtra, Director Emi, Dlohcierekim, DocWatson42, Dr. Whooves, Draceius, Drangon, Drbreznjev, Drmies, DropDeadGorgias, Druid.raul, Dudtz, E23, EEMIV, Easphi, Edhoprima, Effer, Egg plant, Eionm, Eliashedberg, Eloquence, Emperorbma, Enormousdude, Esw01407, Everdawn, FLengyel, Fagarasan, Falcofire, Falcorian, Fastfission, Favonian, Ferengi, Filceolaire, Flayer, Fluffernutter, Flyguy649, Furqan123456789, Fyyer, GCarty, Gail, Gene Nygaard, Georgewilliamherbert, Gesalbte, Gilisa, Gits (Neo), Glane23, Go ahead punk, GraemeLeggett, Graham87, Great.constantine1, GreatWhiteNortherner, GregLoutsenko, GregorB, Greyengine5, Greyhood, Guppie, Gwernol, HP83, HaeB, Hamtechperson, Harman malhotra, Harsha363, Haxxor09, Hdt83, Hede2000, Hektor, Helium4, HeteroZellous, Hibernian, Hobartimus, Hollowman512, Hoover889, Husnock, Huzefa Saifee, Iceman444k, Iknowyourider, Illythr, Imjustmatthew, Ioeth, Iridescent, Isoxyl, IworkforNASA, Ixfd64, J.delanoy, JJJJust, JRM, Jaavaaguru, Jackaranga, Jan Gadimzadeh, Jasgrider, Jeltz, Jenova20, Jhf, Jiy, Jksdguiwaetrwue, Joema, John, JohnDelano, JohnOwens, Joman726, Joseph Solis in Australia, Joshbaumgartner, Jusdafax, Kaakg, Kal-El, KaranJ, Karanrawat21, Ke4roh, Ketiltrout, KidA424, Kinaro, King Zebu, Kingpin13, Kiore, Kirk2, Koavf, Kralizec!, Kubanczyk, KurtRaschke, Kwamikagami, LOL, Lamjus, Leeveraction, LeilaniLad, Li-sung, Lightmouse, LovesMacs, MARQUIS111, MBK004, MUSICAL, MaeseLeon, MagisterLudi, Manishuvits, Marcika, Marcus Qwertyus, Mark Lincoln, Marqueed, Matthew Proctor, Maury Markowitz, Mav, Maxim K, Meeowow, Mejor Los Indios, Melloss, Michael Hardy, Midgetman433, Mieciu K, Mike Rosoft, Missile expert, Mm11, Morven, Mr Fu.ck You, Mrt3366, Mrziggy5000, Muad, Mulad, Muneer2908, Muriel Gottrop, Mythbuster2010, N2e, N328KF, N5iln, NATO UNIT 3, Nabokov, NameIsRon, Nat682, Navraj ghataura, NawlinWiki, Neelix, Neutrality, Nicholasink, Nirvana888, NorseOdin, Nx, Obradovic Goran, Ohconfucius, OlEnglish, Ordinaterr, Ortcutt, Otthgr, Ours, Ovesen, Oxfordwang, Pakzind, PanagosTheOther, PaoloMarcenaro, Passargea, Patrick, Pauli133, PepijnvdG, PhilHibbs, Ploughshares, Pmsyyz, Poached Turkey, PoccilScript, Poliphile, Porckmaster, Pstakem, Pstudier, Pwnage8, Pwrproretaf07, Qazmlp1029, RA0808, RVJ, RajatKansal, Rakopa, Rama, Rangoon11, Raryel, Raza0007, RedWolf, Redefinecool, RevolverOcelotX, Rianamit, Rich Farmbrough, RickK, Rjb uk, Rjd0060, Rjwilmsi, Rlandmann, Rmhermen, Roadrunner, Robert Skyhawk, Rockvee, Rogerd, Rohit spas12, Ronhjones, Rotblats09, RoyBoy, Rupertslander, Russavia, Rwboa22, Rwendland, S trinitrotoluene, S3000, SG, ST47, Sardanaphalus, Scetoaux, Sdsds, Segregator236, Sephirothis666, Sergeyy, ShaunMacPherson, Shriram, Sidkoode, Signsolid, Simetrical, SimranjeetSingh2507, Skipperboi11, Slaporte, Slowking Man, Soakologist, Somatrix, Sp33dyphil, Srikarkashyap, Srjwalker, Stan Shebs, Stanistani, Subsurd, Superm401, Supertask, THE CAKE IS ALIVE, TTE, Tabletop, Tangerineduel, Tavrian, Tbhotch, Tdadamemd, Tec15, Tempshill, Thatpage1798, Thaurisil, The Anome, The Nut, The Rogue Penguin, The Thing That Should Not Be, Thingg, Thomas Larsen, Thumperward, Tim Pritlove, Tkynerd, ToMega666, Toby Douglass, Tom Paine, TomStar81, Tonsitem, Tourbillon, Towel401, Trelvis, Trevor MacInnis, Trongphu, TwoOneTwo, Txinviolet, Uncle Dick, Uris, Usergreatpower, Vedran8080, Versus22, Victor-ny, Victory in Germany, Videogamer22, Vilcxjo, Vineetmenon, Vipin3000, Vistaindia, Vivekrattan, Voltzz, Wafulz, Walle83, Wasbeer, Wavelength, Wayfarer, WikipedianProlific, Will Beback, WriterHound, Wxstorm, Xav71176, Xaveq, Yappy2bhere, Yin61289, Youknowthatoneguy, Zedla, ZeroOne, रोहित रावत, 789 anonymous edits

Strategic_Air_Command *Source*: http://en.wikipedia.org/w/index.php?title=Strategic_Air_Command *Contributors*: 3973cds, 4wajzkd02, Alan.ca, Ansible, Anyeverybody, Anynobody, Apoc2400, Arcadie, ArgentLA, Aspie1, Bahamut0013, Banjodog, Bbpen, Behun, Ben Ben, Bensin, Benstown, Bert Schlossberg, Big D Cowboy, BilCat, Binksternet, Buckboard, Buckshot06, Buffs, Bushcarrot, Bwmoll3, Caknuck, CambridgeBayWeather, Carsc, Centrx, Chadd1227, Chris Globe, ClaudeMuncey, Clngre, Cmp122688, Conelrad79, Connormah, ConradPino, Conversion script, Crowish, D Monack, Dcfowler1, Dennis Brown, DevastatorIIC, Dilbert2147, DocWatson42, Dodgerblue777, Dunc0029, Dunott, Dwtpa97, Enigmaman, Ericg, Fanra, Fastfission, Fl295, Fleminra, FloK, Flyguy649, Fratrep, Frijole, Gail, Georgia guy, Glacier109, Gnoitall, Graham87, Ground Zero, Harmil, Hephaestos, Hildenja, Hmains, Hohum, Howsa12, Humbaband, Hunterd, Husnock, Hydrargyrum, IVinshe, Indrian, Isthisthingworking, James Seneca, Jaraalbe, Jerzy, Jfitts, John, Joseph Dwayne, Jtir, Juliancolton, Jusdafax, Justme89, Karl Dickman, Ketiltrout, Khazar, Kingpin13, Klemen Kocjancic, KristoferM, Ktr101, Kwamikagami, LWF, Larry Sanger, Larry_Sanger, Logawi, LorenzoB, Los688, Lukifer, MBK004, Mais oui!, Mark den Hooghe, MartinHarper, Maury Markowitz, MeekSaffron, Mike Dill, Moormand, Moriori, N328KF, NDCompuGeek, NRW, NawlinWiki, Ndunruh, Neutrality, Nichenbach, Ninly, Nobunaga24, Olivier, Pascal666, Patrick, PaulHanson, Pawyilee, Peripitus, Petri Krohn, Phongn, Pilotguy, Plasticup, Pmsyyz, R'n'B, R. E. Mixer, RHaworth, RJASE1, RadioBroadcast, Redsox00002, Reedmalloy, Retiono Virginian, Rich Farmbrough, Rlandmann, Rogerd, Ron Mixer, Ronald E. Mixer, SAC64, SamBlob, Scatteredpixels, Sj, Skeet Shooter, Sosodank, Spellmaster, Starcheerspeaksnewslostwars, Surv1v4l1st, Syrthiss, Tdrss, Tempodivalse, Tempshill, TimShell, UnitedStatesian, Urhixidur, V Brian Zurita, Vegaswikian, Versus22, Woolhiser, Xyzzyplugh, Zeno Gantner, 359 anonymous edits

United_States_Strategic_Command *Source*: http://en.wikipedia.org/w/index.php?title=United_States_Strategic_Command *Contributors*: 5-HT8, A.R., Achmelvic, Alphachimp, Amcaja, ArCgon, Asten77, AzureCitizen, Bahamut0013, Behun, Beland, BilCat, Buckshot06, C7L5N9, Caltas, Claudevsq, Connormah, Cornellrockey, Courcelles, Crosbiesmith, D6, DJ Clayworth, DangApricot, Ddelestrac, DenverApplehans, Devnull17, Djbauch, DocWatson42, EagleWSO, Fastfission, Florian Adler, Fnlayson, FrozenPurpleCube, Gaius Cornelius, Gbbinning, Glacier109, Goldom, HJ32, Harry491, Hcobb, Hmains, Ida Shaw, JCO312, JamesGraybeal, Jigen III, Jogrkim, Jonverve, Jpm2112, JustAGal, Ken Gallager, Kent Wang, Klemen Kocjancic, Kumioko (renamed), LanceBarber, Lightmouse, LilHelpa, MCG, MLWilson, Mais oui!, Mandarax, Marcd30319, Mark83, Mdnort, Mhustoft, MrDolomite, Ndunruh, Neovu79, Neutrality, Night Gyr, Nobunaga24, Nohomers48, Ohconfucius, Pgrote, Pmsyyz, Portkent, Preuninger, Rejectwater, Retanollo, RicJac, Rich Farmbrough, Rougher07, Saga City, SeanFromIT, Seitzd, SimonP, Smug Irony, Stevertigo, Student7, Sunray, Syrthiss, Tdrss, Thesmothete, Tim!, Tomtom, Tourbillon, USSTRATCOM PAO, Uv1234, Walker9010, Warrenfish, Woogers, Xanzzibar, 122 anonymous edits

Retroactive_continuity *Source*: http://en.wikipedia.org/w/index.php?title=Retroactive_continuity *Contributors*: 23skidoo, 24.82.9.xxx, ABCxyz, Adam Conover, Adam Keller, Admiral Roo, Ahoerstemeier, AlbinoChocobo, Alerante, Alex Weitzman, Alientraveller, AlistairMcMillan, Aliza250, Alton.arts, Amalas, Ambientlight, AmuroNT1, AnalysisInfinitorum, Andreas Kaganov, Andyroo316, Antaeus Feldspar, Anthony Appleyard, Apostrophe, Aprock, Arjayay, Asa01, Astronautics, Atlantima, Atropos, Austinfidel, Axl, Azrael81, BabySinclair, Basique, Baylink, Bbpen, Bdve, BenBaker, Bill37212, Blanche of King's Lynn, Blissyb, Bloodshedder, Bluesykillerhorn, Bookrat, Boxclocke, Brian Kendig, Bryan Derksen, Bssc81, Buckyboy28, Bulmabriefs144, Bxj, CFLeon, CajunGypsy, Caltrop, Cameron Scott, CanadianCaesar, Cananian, Captain Yesterday, CatMarieS, CatherS, Cburnett, Ccsccs7, Centrx, Charles Matthews, Cheesy123456789, Chemako0606, ChrisGriswold, Civil Engineer III, Clampton, Clegs, Closedmouth, Colliric, Colonies Chris, Comicist, Conan-san, ConradKilroy, Conti, Conversion script, Cooksey, Corby, Corixidae, Courcelles, CovenantD, Croctotheface, CrossEyed7, Cvkline, CyberGhostface, Cybertooth85, Cyhwuhx, D. F. Schmidt, DCGeist, DRosenbach, DT29, Dale Arnett, Daniel Lawrence, Darkwind, Dblack88, DcPimp, Derekcohen, Derktar, Destroyer of evil, Deuxhero, Dippit, Discographer, DividedByNegativeZero, Dmlandfair, Doczilla, DonQuixote, Doom127, Download, Dr. R.K.Z, DragonflySixtyseven, Dravecky, Dsreyn, Dstumme, Duggy 1138, Dustinasby, EamonnPKeane, Efb18, Egolub, Ellsworth, Elvenscout742, Elysdir, Emperor, Epbr123, Ergative rlt, Erianna, ErrantX, Esperant, Esrever, Fabulous Creature, Fadookie, Farlstendoiro, Fastfission, Feedmecereal, Fhb3, Fourthords, Fratrep, Fred8615, Ftld, Furrymoose, GaidinBDJ, Gary D, Gaz C, GeneralDuke, George100, Gogo Dodo, Golem866, Gonzalo84, Gothamgazette, Gouryella, GracieLizzie, Graham87, Grand Dizzy, Grandpafootsoldier, Gregory j, Guru Larry, HMP22, Haoie, HeartBurn Kid, HelenKMarks, Hes Nikke, Hiroshi-br, Hmains, HunterZ, I want a big stereo, Icarus3, IceHunter, InShaneee, Indium, Intractable, Iodyne, Iron Sun 254, Itai, J Greb, JIP, JPX7, JSmith9579, Jabberwoch, JamesMLane, JasonAQuest, Jedibob5, Jetfire85, Jhsounds, Jhz, Jiy, Joeyconnick, JohnDBuell, JonStrines, JonathanDP81, JonathanNil, Jonny2x4, Josh Cherry, Josiah Rowe, Joylock, Julian Grybowski, Julu11, Jungleonvinyl, JustPhil, Kchisholl1970, Ken Arromdee, Kent Wang, Klenod, Koyaanis Qatsi, Ksofen666, L33tminion, LDEJRuff, LOwen, LeCire, Leocomix, Lg16spears, LilHelpa, Lokicarbis, LonerXL, Lowellian, Luigifan, Luis Dantas, Luna Santin, Lysdexia, M.C. Brown Shoes, MCB, Macker, Macktheknifeau, Magioladitis, MakeRocketGoNow, Marcus Brute, MarkSutton, Master Deusoma, Master Thief Garrett, Mateo SA, Mattbuck, Mboverload, McGeddon, Mccaffry, MegX, Mel Etitis, Melaen, Mezaco, Michael Rawdon, Michael Reiter, Mindbleach, Molly-in-md, Mordantkitten, Morrigan Targaryen, Mr. ATOZ, MrBook, Mtconnell, MuzikJunky, Myleslong, Naddy, Nalvage, Nautilator, Neelix, Neilbeach, Nidonocu, Nightscream, Nightscreamnovi, Nintendo Maximus, Niteowlneils, Nnordlinger, Noclevername, Norm mit, OGRastamon, OJFC, Olivier, Ophois, Opticality, Otto4711, Pacholeknbnj, Paddles, Pakaran, Palfrey, Patricia Ullman, Patrick, Paul A, Paul August, Pax:Vobiscum, Pdc, PenguiN42, Peter Delmonte, Photonh2o, Phthoggos, Phydend, Pictureuploader, Pikawil, PlasmaDragon, Platypus222, Poccil, Poiuyt Man, Populus, Postdlf, Psyk0, Psykhaotic, Ptcamn, Pthag, Quicksilvre, QuizzicalBee, RK, Rapturerocks, Rattlerbrat, Raymondluxuryacht, Rbarreira, Reach Out to the Truth, Red Scharlach, RedWolf, RexNL, Rezdave, Rich Farmbrough, Richard75, Rillian, Riumplus, Rlquall, Roadrunner, RobertCMWV1974, Robertd, Rockhopper10r, Rompe, Rorschach567, Rpike20625, Ryulong, SJennings, Sailorptah, Salamurai, SaliereTheFish, Sanchom, SandChigger, Sandor Clegane, Sashal, SchuminWeb, Scottandrewhutchins, Scumbag, Septegram, Sesamekinjiru, SexyIrishLeprechaun, ShaleZero, ShelfSkewed, SidP, Sietse Snel, Sikon, Sinalese, Sloman, Sonett72, Spidey104, Sproutviewer, SquarePeg, Staecker, Starfarmer, Stdjsb25, Stefanomione, Stellis, StephenP98, Stillnotelf, Stirling Newberry, Stoshmaster, Sundevil4life, Syndicate, SynjoDeonecros, Sysy, TAnthony, Tamfang, Tassedethe, Template namespace initialisation script, Ten-pint, Tenchichan, That Guy, From That Show!, The Anome, The Man in Question, The Original Threepwood, The Wookieepedian, TheRanger, TheRunba, Therealorion, Thesis4Eva, Thu, TimBentley, Timon, Timrollpickering, Timwi, Toba, Tokachu, TopAce, Toquinha, Tpbradbury, Trampikey, Traxs7, Tregoweth, Trvsdrlng, Trystan, Turnstep, Tverbeek, Tvoz, Twinxor, U-Mos, Uliwitness, Ultraexactzz, UnitedStatesian, Urhixidur, Varlaam, Vendettax, Vzbs34, WCityMike, Wereon, WhisperToMe, Wickethewok, Wik, Wikipediatrix, Wkitech, Wl219, Wwoods, Xcarex, Yar Kramer, YoungFreud, Zentinel, Zippanova, Zythe, Ὁ οἶστρος, 732 anonymous edits

AN/FSQ-31V *Source*: http://en.wikipedia.org/w/index.php?title=AN/FSQ-31V *Contributors*: Auntof6, ChardingLLNL, ChuckHarding53, Dual Freq, Gouru, Guy Harris, Ketiltrout, Kumioko (renamed), NDCompuGeek, Ndunruh, Neelix, O keyes, PigFlu Oink, RTC, Rjwilmsi, Thadius856, Thunderbird2, 6 anonymous edits

Curtis_LeMay *Source*: http://en.wikipedia.org/w/index.php?title=Curtis_LeMay *Contributors*: A. Carty, AWasteBasket, Abraham, B.S., Adashiel, Ahoerstemeier, Alecmconroy, Alex Avalon, Alex890, Alhutch, All Hallow's Wraith, Andrwsc, AnmaFinotera, Antandrus, AntonyZ, ArglebargleIV, Arix13, Arla, Ashdog137, Askari Mark, Atlanticcoast63, Att109, Avriette, Badbeagle, Bahamut0013, Bazdmeg14, Bbsrock, Beanbatch, Bender235, Bethpage89, Beyond My Ken, Billy Hathorn, Bilsonius, Binksternet, Biruitorul, Bnguyen, BobbyG86, Brandon, BrandonTR, BrianY, Brothernight, Buckboard, Buckshot06, Buffs, Butseriouslyfolks, Bwmoll3, Callidior, Calton, Canglesea, CanisRufus, CardinalDan, Carl Logan, Carthago delenda est, Chris the speller, Chrislk02, Connormah, Crimson stranger, Cycleman63, Cyrano 21, D6, DS1953, Dandvsp, Daniel Quinlan, Darth Kalwejt, DarthRad, DavidA, Daxmac, Dellant, Destroyer2, Deville, Dhanig, Dhruvhemmady, DocWatson42, Donfbreed, Donreed, Download, DreamGuy, Drilnoth, Dstub, Dudeman5685, Dumelow, E-Dogg, EHDI5YS, EHRice, ERcheck, Edward, Ekotkie, Erik9, Evans1982, Everyking, Ewildered, Explainer, FLJuJitsu, Fastfission, Flyboy03, Fnlayson, Folks at 137, Fredrik, Friedlad, Ft93110, Gadfium, Gary123, Ged UK, Giraffedata, Glacier109, Glennwells, Good Olfactory, GoodDay, Googie man, Grant65, GreatOrangePumpkin, GregorB, Grenavitar, Ground Zero, Guildtheeli, Gunter, Hamdrinker, Hammersoft, Handicapper, Havermayer, Hcheney, Hcobb, Hephaestos, Hidudewasup, Hmains, Hollywood Editor, Hoponpop69, Huangdi, IAC-62, ITOD, Iamlondon, Iloveandrea, Imaginaryoctopus, Indech, Infrogmation, Innapoy, Ipankonin, IronMaidenRocks, ItemCo16527, JCO312, JD554, JE1977, Jacob1207, Jake Wartenberg, JayJasper, Jef-Infojef, Jelson25, Jeremy Bentham, Jigen III, Jimipuppet, Jni, Jonas Mur, Jonathan.s.kt, Joseph Solis in Australia, Joshmaul, Jufemaiz, Jusdafax, Jwillbur, KD5TVI, Karl gregory jones, Kchishol1970, Kd5npf, Kernel Saunters, Khoikhoi, Klemen Kocjancic, Koavf, Kobalt64, Kransky, Krazychris81, Kubanczyk, Kumioko (renamed), Kurihaya, Kwertii, L., Larryincinci, Learner001, Lestatdelc, Leszek Jańczuk, Lightmouse, Littlemo, Lommer, LorenzoB, Lseltzer, MBK004, MK2, MachoCarioca, Marhar, Marine 69-71, Mark Heiden, Mark Sublette, MartinDK, Mass147, Mattnad, Maverick9711, Mdnavman, MichaelGood, Miette49, Mild Bill Hiccup, Mintleaf, Miss Madeline, MisterJayEm, Modest Genius, Moheroy, Moncrief, Motoraptor, Movedgood, Movementarian, MrDolomite, Mrb712, Mtracy9, N328KF, Nanshu, Narpull, Ndunruh, Neovu79, Niceguyedc, Nick Number, NightMonkey, Nixdorf, Nobunaga24, Nono64, Num1dgen, Number29, Nymuseum, Nyr14, OberRanks, Oddharmonic, P1340, PRehse, Padddy5, Part Deux, PaulHanson, PaulinSaudi, Pearle, PedanticallySpeaking, Percommode, PhantomWSO, PhilKnight, Philip Baird Shearer, Piccadilly, Pmanderson, Pooder12, Protonk, Qiitxx, Qworty, RATHED, RSStockdale, RandomStringOfCharacters, Ravenswing, Recognizance, Reedmalloy, Reenem, Rettetast, Revth, RexNL, RicJac, Rich Farmbrough, Richard Weil, Rjwilmsi, Rkevins, Rlfielder, Robespierre2, Rogerd, Ryan Roos, SchmuckyTheCat, ScottJ, SeanO, Seattle Skier, Shiai, Skubasteve834, Smurrayinchester, Snowmanradio, Sonicology, Spot87, Stan Shebs, Steve8675309, SteveSims, Studerby, Sully, T.E. Goodwin, TBHecht, TFBCT1, TL36, Tabletop, Target for Today, Tassedethe, Tenmei, Thaurisil, The Utahraptor, The wub, Themoodyblue, Thismightbezach, Tom, Tommyt, TonyBushido, Topbanana, Tothebarricades.tk, Trekie8472, Trojancowboy, Twinxor, Ulflarsen, UtherSRG, Vapour, Viriditas, Voronwae, Waacstats, Walter Dufresne, Walton One, Warren Kozak, Warwick555, Wellreadone, Welsh, Wenatcheeman, Wiki alf, WikiDan61, WikigeojeeK, Wikiuser1239, Wolbo, Woohookitty, Worldruler20, Xchanter, Xdamr, XiaoweiT, Ylee, Zoe, , 403 anonymous edits

Image Sources, Licenses and Contributors

Image:SAC Automated Command and Control System.png *Source*: http://en.wikipedia.org/w/index.php?title=File:SAC_Automated_Command_and_Control_System.png *License*: unknown *Contributors*: ChardingLLNL

File: Command Data Buffer configuration.png *Source*: http://en.wikipedia.org/w/index.php?title=File:Command_Data_Buffer_configuration.png *License*: unknown *Contributors*: United States Air Force

Image:Titan II launch.jpg *Source*: http://en.wikipedia.org/w/index.php?title=File:Titan_II_launch.jpg *License*: unknown *Contributors*: Craigboy, FlickreviewR, Lymantria, SatuSuro, WDGraham

Image:minuteman3launch.jpg *Source*: http://en.wikipedia.org/w/index.php?title=File:Minuteman3launch.jpg *License*: unknown *Contributors*: User:Edbrown05

File:Dnepr rocket lift-off 1.jpg *Source*: http://en.wikipedia.org/w/index.php?title=File:Dnepr_rocket_lift-off_1.jpg *License*: unknown *Contributors*: User:NH2501/Kosmotras

File:Semyorka 8K71.svg *Source*: http://en.wikipedia.org/w/index.php?title=File:Semyorka_8K71.svg *License*: unknown *Contributors*: WDGraham

File:USAF ICBM and NASA Launch Vehicle Flight Test Successes and Failures (highlighted).png *Source*: http://en.wikipedia.org/w/index.php?title=File:USAF_ICBM_and_NASA_Launch_Vehicle_Flight_Test_Successes_and_Failures_(highlighted).png *License*: unknown *Contributors*: Tdadamemd

File:Peacekeeper missile after silo launch.jpg *Source*: http://en.wikipedia.org/w/index.php?title=File:Peacekeeper_missile_after_silo_launch.jpg *License*: unknown *Contributors*: Cobatfor, Ibonzer, Uwe W., 1 anonymous edits

Image:TridentMissileSystem.png *Source*: http://en.wikipedia.org/w/index.php?title=File:TridentMissileSystem.png *License*: unknown *Contributors*: Original uploader was WikipedianProlific at en.wikipediabr/> (Original text :)

Image:Peacekeeper-missile-testing.jpg *Source*: http://en.wikipedia.org/w/index.php?title=File:Peacekeeper-missile-testing.jpg *License*: unknown *Contributors*: Original uploader was Solipsist at en.wikipedia

File:SS-24-DIA.jpg *Source*: http://en.wikipedia.org/w/index.php?title=File:SS-24-DIA.jpg *License*: unknown *Contributors*: Edward L. Cooper

Image:SAC Shield.svg *Source*: http://en.wikipedia.org/w/index.php?title=File:SAC_Shield.svg *License*: unknown *Contributors*: United States Air Force

File:Carswell Bomber Static Display.jpg *Source*: http://en.wikipedia.org/w/index.php?title=File:Carswell_Bomber_Static_Display.jpg *License*: unknown *Contributors*: United States Air Force

File:SAC 1947 OC.gif *Source*: http://en.wikipedia.org/w/index.php?title=File:SAC_1947_OC.gif *License*: unknown *Contributors*: R. E. Mixer

File:Sac194-patch.png *Source*: http://en.wikipedia.org/w/index.php?title=File:Sac194-patch.png *License*: unknown *Contributors*: Original uploader was Ron Mixer at en.wikipedia

Image:B52D 56-0687.JPG *Source*: http://en.wikipedia.org/w/index.php?title=File:B52D_56-0687.JPG *License*: unknown *Contributors*: Fl295

Image:Titan2 color silo.jpg *Source*: http://en.wikipedia.org/w/index.php?title=File:Titan2_color_silo.jpg *License*: unknown *Contributors*: Badzil, Balmung0731, Bricktop, Bwmoll3, Cobatfor, Denniss, Dual Freq, Fastfission, High Contrast

File:USSTRATCOM.svg *Source*: http://en.wikipedia.org/w/index.php?title=File:USSTRATCOM.svg *License*: unknown *Contributors*: Not specified

File:LeMaybldg.jpg *Source*: http://en.wikipedia.org/w/index.php?title=File:LeMaybldg.jpg *License*: unknown *Contributors*: Monkeybait, Schlendrian, 1 anonymous edits

File:GEN George L Butler.jpg *Source*: http://en.wikipedia.org/w/index.php?title=File:GEN_George_L_Butler.jpg *License*: unknown *Contributors*: USAF

File:Henry G Chiles.jpg *Source*: http://en.wikipedia.org/w/index.php?title=File:Henry_G_Chiles.jpg *License*: unknown *Contributors*: United States Navy

File:Eugene E Habiger.jpg *Source*: http://en.wikipedia.org/w/index.php?title=File:Eugene_E_Habiger.jpg *License*: unknown *Contributors*: Original uploader was Nobunaga24 at en.wikipedia

File:Richard W Mies.jpg *Source*: http://en.wikipedia.org/w/index.php?title=File:Richard_W_Mies.jpg *License*: unknown *Contributors*: Original uploader was Nobunaga24 at en.wikipedia

File:James o ellis.jpg *Source*: http://en.wikipedia.org/w/index.php?title=File:James_o_ellis.jpg *License*: unknown *Contributors*: Original uploader was Nobunaga24 at en.wikipedia

File:James E. Cartwright.jpg *Source*: http://en.wikipedia.org/w/index.php?title=File:James_E._Cartwright.jpg *License*: unknown *Contributors*: Anathema, FieldMarine, Kelly, Mattes, Nobunaga24, Schimmelreiter, 3 anonymous edits

File:C. Robert Kehler 2007.jpg *Source*: http://en.wikipedia.org/w/index.php?title=File:C._Robert_Kehler_2007.jpg *License*: unknown *Contributors*: GrummelJS, Nobunaga24

File:Kevin P. Chilton.jpg *Source*: http://en.wikipedia.org/w/index.php?title=File:Kevin_P._Chilton.jpg *License*: unknown *Contributors*: Claudevsq, GrummelJS, Nobunaga24, Väsk, 1 anonymous edits

File:USAF General C. Robert Kehler.jpg *Source*: http://en.wikipedia.org/w/index.php?title=File:USAF_General_C._Robert_Kehler.jpg *License*: unknown *Contributors*: United States Department of Defense

File:Curtis LeMay (USAF).jpg *Source*: http://en.wikipedia.org/w/index.php?title=File:Curtis_LeMay_(USAF).jpg *License*: unknown *Contributors*: Fastfission, Felix Stember, Historicair, Nobunaga24, Rklawton, SoLando

File:Flag of the United States.svg *Source*: http://en.wikipedia.org/w/index.php?title=File:Flag_of_the_United_States.svg *License*: unknown *Contributors*: Anomie

File:Seal of the US Air Force.svg *Source*: http://en.wikipedia.org/w/index.php?title=File:Seal_of_the_US_Air_Force.svg *License*: unknown *Contributors*: US Army Institute Of Heraldry

File:US-O10 insignia.svg *Source*: http://en.wikipedia.org/w/index.php?title=File:US-O10_insignia.svg *License*: unknown *Contributors*: user:Ipankonin

File:B-29s dropping bombs.jpg *Source*: http://en.wikipedia.org/w/index.php?title=File:B-29s_dropping_bombs.jpg *License*: unknown *Contributors*: User:W.wolny

File:Firebombing leaflet.jpg *Source*: http://en.wikipedia.org/w/index.php?title=File:Firebombing_leaflet.jpg *License*: unknown *Contributors*: Fastfission, FayssalF, M9106TB, Nick-D, PMG, Schekinov Alexey Victorovich, Teofilo, Tm, 1 anonymous edits

File:lemay4.jpg *Source*: http://en.wikipedia.org/w/index.php?title=File:Lemay4.jpg *License*: unknown *Contributors*: USAF

File:COMMAND PILOT WINGS.png *Source*: http://en.wikipedia.org/w/index.php?title=File:COMMAND_PILOT_WINGS.png *License*: unknown *Contributors*: Dandvsp, Ed!, Hawkeye7, SGT141, 2 anonymous edits

File:Distinguished Service Cross ribbon.svg *Source*: http://en.wikipedia.org/w/index.php?title=File:Distinguished_Service_Cross_ribbon.svg *License*: unknown *Contributors*: user:Ipankonin

Image:Bronze oakleaf-3d.svg *Source*: http://en.wikipedia.org/w/index.php?title=File:Bronze_oakleaf-3d.svg *License*: unknown *Contributors*: lestatdelc

Image:Distinguished Service Medal ribbon.svg *Source*: http://en.wikipedia.org/w/index.php?title=File:Distinguished_Service_Medal_ribbon.svg *License*: unknown *Contributors*: user:Ipankonin

File:Silver Star ribbon.svg *Source*: http://en.wikipedia.org/w/index.php?title=File:Silver_Star_ribbon.svg *License*: unknown *Contributors*: user:Ipankonin

Image:Distinguished Flying Cross ribbon.svg *Source*: http://en.wikipedia.org/w/index.php?title=File:Distinguished_Flying_Cross_ribbon.svg *License*: unknown *Contributors*: user:Ipankonin

Image:Air Medal ribbon.svg *Source*: http://en.wikipedia.org/w/index.php?title=File:Air_Medal_ribbon.svg *License*: unknown *Contributors*: user:Ipankonin

Image:AF Presidential Unit Citation Ribbon.png *Source*: http://en.wikipedia.org/w/index.php?title=File:AF_Presidential_Unit_Citation_Ribbon.png *License*: unknown *Contributors*: User:Dandvsp

File:American Defense Service ribbon.svg *Source*: http://en.wikipedia.org/w/index.php?title=File:American_Defense_Service_ribbon.svg *License*: unknown *Contributors*: user:Ipankonin

File:American Campaign Medal ribbon.svg *Source*: http://en.wikipedia.org/w/index.php?title=File:American_Campaign_Medal_ribbon.svg *License*: unknown *Contributors*: user:Ipankonin

Image:Bronze-service-star-3d.png *Source*: http://en.wikipedia.org/w/index.php?title=File:Bronze-service-star-3d.png *License*: unknown *Contributors*: User:Lestatdelc

Image:European-African-Middle Eastern Campaign ribbon.svg *Source*: http://en.wikipedia.org/w/index.php?title=File:European-African-Middle_Eastern_Campaign_ribbon.svg *License*: unknown *Contributors*: user:Ipankonin

Image:Asiatic-Pacific Campaign ribbon.svg *Source*: http://en.wikipedia.org/w/index.php?title=File:Asiatic-Pacific_Campaign_ribbon.svg *License*: unknown *Contributors*: user:Ipankonin

File:World War II Victory Medal ribbon.svg *Source*: http://en.wikipedia.org/w/index.php?title=File:World_War_II_Victory_Medal_ribbon.svg *License*: unknown *Contributors*: user:Ipankonin

File:AirliftDev.jpg *Source*: http://en.wikipedia.org/w/index.php?title=File:AirliftDev.jpg *License*: unknown *Contributors*: FSII, FieldMarine

File:Medal for Humane Action ribbon.svg *Source*: http://en.wikipedia.org/w/index.php?title=File:Medal_for_Humane_Action_ribbon.svg *License*: unknown *Contributors*: user:Ipankonin

File:National Defense Service Medal ribbon.svg *Source*: http://en.wikipedia.org/w/index.php?title=File:National_Defense_Service_Medal_ribbon.svg *License*: unknown *Contributors*: user:Ipankonin

Image:Silver oakleaf-3d.svg *Source*: http://en.wikipedia.org/w/index.php?title=File:Silver_oakleaf-3d.svg *License*: unknown *Contributors*: User:Lestatdelc

Image:Air Force Longevity Service ribbon.svg *Source*: http://en.wikipedia.org/w/index.php?title=File:Air_Force_Longevity_Service_ribbon.svg *License*: unknown *Contributors*: user:Ipankonin

File:Uk dfc rib.png *Source*: http://en.wikipedia.org/w/index.php?title=File:Uk_dfc_rib.png *License*: unknown *Contributors*: AusTerrapin, J a1, Mboro

File:Croix de guerre 1939-1945 with palm.jpg *Source*: http://en.wikipedia.org/w/index.php?title=File:Croix_de_guerre_1939-1945_with_palm.jpg *License*: unknown *Contributors*: User:EHDI5YS

File:Oorlogskruis with Palm.jpg *Source*: http://en.wikipedia.org/w/index.php?title=File:Oorlogskruis_with_Palm.jpg *License*: unknown *Contributors*: User:EHDI5YS

File:JPN Kyokujitsu-sho 1Class BAR.svg *Source*: http://en.wikipedia.org/w/index.php?title=File:JPN_Kyokujitsu-sho_1Class_BAR.svg *License*: unknown *Contributors*: User:Mboro

File:Order of the Southern Cross Grand Collar Ribbon.png *Source*: http://en.wikipedia.org/w/index.php?title=File:Order_of_the_Southern_Cross_Grand_Collar_Ribbon.png *License*: unknown *Contributors*: Dandvsp (talk). Original uploader was Dandvsp at en.wikipedia

File:Santos-Dumont Medal of Merit Ribbon.png *Source*: http://en.wikipedia.org/w/index.php?title=File:Santos-Dumont_Medal_of_Merit_Ribbon.png *License*: unknown *Contributors*: Dandvsp (talk). Original uploader was Dandvsp at en.wikipedia

File:Commander of the Order of Ouissam Alaouite ribbon.png *Source*: http://en.wikipedia.org/w/index.php?title=File:Commander_of_the_Order_of_Ouissam_Alaouite_ribbon.png *License*: unknown *Contributors*: Dandvsp (talk). Original uploader was Dandvsp at en.wikipedia

Image:LeMaybldg.jpg *Source*: http://en.wikipedia.org/w/index.php?title=File:LeMaybldg.jpg *License*: unknown *Contributors*: Monkeybait, Schlendrian, 1 anonymous edits

Printed by Books on Demand GmbH, Norderstedt / Germany